Mathematics Study Resources

Band 2

Series Editors

Kolja Knauer, Departament de Matemàtiques Informàtic,
Universitat de Barcelona, Barcelona, Barcelona, Spain
Elijah Liflyand, Department of Mathematics, Bar-Ilan University,
Ramat-Gan, Israel

This series comprises direct translations of successful foreign language titles, especially from the German language.
Powered by advances in automated translation, these books draw on global teaching excellence to provide students and lecturers with diverse materials for teaching and study.

Ehrhard Behrends

Tilings of the Plane

From Escher via Möbius to Penrose

 Springer

Ehrhard Behrends
Fachbereich Mathematik und Informatik
Freie Universität Berlin
Berlin, Germany

ISSN 2731-3824   ISSN 2731-3832   (electronic)
Mathematics Study Resources
ISBN 978-3-658-38809-6   ISBN 978-3-658-38810-2   (eBook)
https://doi.org/10.1007/978-3-658-38810-2

Responsible Editor: Iris Ruhmann
This Springer imprint is published by the registered company Springer Fachmedien Wiesbaden GmbH, part of Springer Nature.
The registered company address is: Abraham-Lincoln-Str. 46, 65189 Wiesbaden, Germany

Preface

The present book deals with *three special aspects* of the topic "tilings of the plane", which have an interesting mathematical background. By a tiling is meant a seamless and non-overlapping coverage of the plane, where the individual building blocks arise from a "simple" law of formation - for example by the action of a group of motions.

In the *first part* we consider motions of the plane, which preserve distances and then study objects which are invariant under certain motions: This leads to the concept of *symmetry*. As a very simple example for illustration one could consider the letter "M": If one reflects it at the median line, it merges into itself. Much more interesting are of course examples from architecture (rotational and mirror symmetry) or art: The pictures of the Dutch graphic artist Maurits Cornelis Escher, who was inspired by the patterns in the Alhambra in Granada, are well known and then realized various aspects of symmetry in his graphics masterfully. We will look "over the shoulder" of Escher and derive those mathematical results and construction methods which he was able to find by artistic intuition, without ever having had a mathematical education.

A rosette from the Museum of Applied Arts in Vienna

In *the second part* the topic of symmetry is interpreted from the perspective of function theory. The natural "movements" of the complex number sphere are those which are firstly holomorphic and secondly bijective. They are easy to describe, they have the form $z \mapsto (az + b)/(cz + d)$, where a, b, c, d are complex numbers with $ad - bc \neq 0$. Today they are called *Möbius transformations*. We will classify Möbius transformations and see how they and the groups they generate give rise to interesting tessellations of the plane.

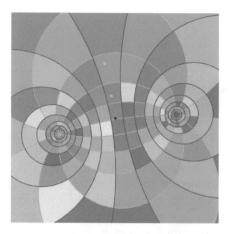

A visualization of a special Möbius transformation

The *third part* is finally dedicated to *Penrose tessellations*. Starting from two simple triangles, where the ratio of their sides is the golden section. If you have an unlimited number of these triangles available and postulate some laying rules, it turns out: You can tessellate the plane in principle in infinitely many different ways with these triangles, but none of these tessellations is periodic, that is, it cannot be brought into itself by a non-trivial translation. This solves the long-standing open problem of whether, in the case of a tessellation with certain building blocks, there must also be a periodic tessellation with these building blocks.

A Penrose tessellation (detail)

The book is very detailed and therefore not only suitable as a template for a lecture or seminar, but also for self-study. Through numerous images, the mathematical facts are visualized, and the readers may be inspired to create a picture à la Escher themselves, to create an attractive image using a group of Möbius transformations, or to try a Penrose tiling.

In the context of proseminars and seminars, I have repeatedly addressed the various aspects of the topic "tilings", and in the winter semester of 2017/2018 there was a lecture announced under the title of this book. In it I received many suggestions from the participants, to whom I would like to express my sincere thanks at this point.

Berlin, Germany   Ehrhard Behrends
2018

Contents

Part III Penrose Tilings

References for Part III

Introduction

My main areas of interest are actually functional analysis and probability theory, but from time to time I have also been interested in the topics of "tilings" and "symmetry". It started several decades ago with a question asked by a friend: "You're a mathematician. How can you actually design your own wallpaper pattern?"

I had never dealt with this before, and I found it fascinating to learn how to classify all the different possible patterns: These are the 17 plane crystallographic groups, which are called "wallpaper groups" in English.

There is also artistically demanding visual material, because in the works of the Dutch graphic artist Maurits Cornelis Escher all possibilities have been exhausted. With regard to the initial question, it was also interesting for me to find out that there is a complete catalogue of manufacturing processes found by Heinrich Heesch, which allows one to try out creative wallpaper design for oneself.

The next impetus came at some point in the first years of the new millennium, when I came across the book "Indra's Pearls—The Vision of Felix Klein" by Mumford, Series and Wright. I know of no other source that encourages one to illustrate demanding mathematics in such an attractive way. I myself like to program and often understand mathematical concepts much better if there is a suitable visualization of the mathematical background[1].

A few years later, my interest in Penrose tessellations was aroused by a newspaper article. Since I am always looking for visually attractive mathematical concepts that can be implemented as part of my activities to bring the fascination of mathematics closer to the interested public, I have worked my way into the topic.

---

[1] The motto of a travel guide could be: "You only see what you know". To make it clearer what I mean, one could—greatly simplifying—change it to "You only know what you see".

© The Author(s), under exclusive license to Springer Fachmedien Wiesbaden GmbH, part of Springer Nature 2022
E. Behrends, *Tilings of the Plane*, Mathematics Study Resources 2,
https://doi.org/10.1007/978-3-658-38810-2_1

We also had quite a few large Penrose tiles made, which were used several times on suitable occasions (Long Night of the Sciences, Mathematics Day, …). There was always a lecture adapted to the level of knowledge of the audience, and then you could try to apply the quite complicated assembly rules à la Penrose yourself.

Beginning of a Penrose tiling (Long Night of the Sciences Berlin, 2008)

Now to the content. *Part I* has the title "Escher seen over the shoulder", and that's how it's meant: We want to understand the mathematics that lies behind tiling of the plane so well that we—in principle—could produce images à la Escher with sufficient artistic talent. To achieve this goal, we proceed in the following steps. First, we take care of basics in *Chapter 1*: What is symmetry? How can one classify the movements of the plane? What symmetry group does a given image have? Which tilings by so-called "fundamental domains" are generated by symmetry groups? In *Chapter 2* we take on characterizations: Which symmetry groups are to be expected if there are no or one- or two-parametric families of translations? We restrict ourselves to *discrete groups*, that is, those for which the movements that lead to symmetries do not "come too close" to each other.

The case in which there are no non-trivial translations is quite simple. There can only be rotations by divisors of 360 degrees and reflections (Leonardo's theorem). More interesting is the characterization of the seven *Frieze groups*, there is a one-dimensional set of translations. Much more involved is the treatment of the case in which there are two linearly independent translations in the group. By careful analysis of the then possible translations it can be shown that there are 17 fundamentally different possibilities, these are the 17 *plane crystal groups*. (Of course, it had to be clarified what "essentially different" means in this context.)

The starting point of these investigations is an important lemma, the *crystallographic restriction:* Possible rotations are only those by integer multiples of

180 degrees, 120 degrees, 90 degrees, 72 degrees and 60 degrees. This drastically reduces the number of cases to be examined.

Characterizations are of great theoretical interest, but one would also like to know how to make practical use of them. In order to give all artistically interested people the opportunity to be practically active themselves, we describe in *Chapter 3* the *Heesch constructions*: In 28 essentially different ways one arrives at tessellations which arise from the plane crystal groups. Preparatory to this, we deal with lattices and nets, which make it possible to carry out something like a combinatorial analysis of the annealing possibilities relevant here.

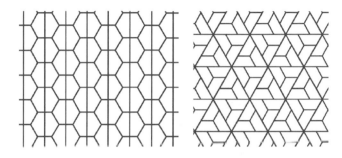

The nets $(4, 4, 3, 3, 3)$ and $(6, 3, 3, 3, 3)$

Here is an illustration of the tenth Heesch construction as an example:

The tiling that belongs to the 10th Heesch construction

In *Part II* we investigate *Möbius transformations and groups of such transformations*. *Chapter 4* begins with a review of some facts related to complex numbers and complex-valued functions. Then Möbius transformations are introduced. The need to study these maps and groups of such maps arose in the effort to better understand multivalued holomorphic functions.

The first important results follow: How are Möbius transformations characterized? How can they be composed of simple transformations? How are circles mapped under Möbius transformations? … This will already give rise to interesting pictures.

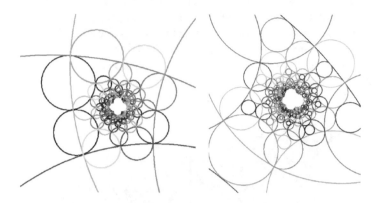

Circles transformed with a Möbius transformation

After an investigation of the possible fixed points of Möbius transformations, a characterization follows: In addition to the identity, there are four essentially different types: parabolic, elliptic, hyperbolic and loxodromic transformations. With the help of many pictures it is illustrated how they can be imagined.

A loxodromic transformation

All of these are necessary preparations in order to begin studying *groups of Möbius transformations* in *Chapter 5*. First, there are those transformations which leave a special subset of the complex plane—such as the upper half-plane—invariant. We take the opportunity for a *digression into hyperbolic geometry*. There, lines and triangles look very unusual, and even the angle sum in a triangle must be derived using a new formula.

Lines and triangles in hyperbolic geometry

We also learn some important properties of the *modular group*, and we generate a well-known tiling by choosing a fundamental domain.

A tiling of $\mathbb{H}$ induced by the modular group

After that we study *Schottky groups*. These are groups with two generators, for which the effects of the occurring transformations can still be described relatively simply. As groups they are free groups, and as the smallest invariant set one obtains the Cantor discontinuum. It becomes particularly interesting when the Schottky circles touch, then there are interesting connected limit sets. The mathematical background is interesting: One has to ensure that the commutator of the generators is a parabolic transformation.

Then we have enough interesting examples for illustration, to turn to the general case: What can be said about *general Kleinian groups*? It turns out that the isometric circles belonging to the transformations play a special role, so that structural statements about the limit set and possible fundamental domains can be proved.

Then we take up the problem of parabolic commutators again: How can one find examples? A fundamental role is played by the *Markov identity*, with which it is possible to give many families of examples. For the limit sets interesting fractal structures then result.

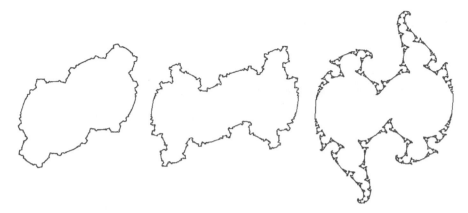

Some limit sets

Part III is dedicated to the study of *Penrose tiling*. *Chapter 6* begins with a clarification of the problem and a short historical overview. Then we introduce the *building blocks*, they are triangles, in whose definition the golden section plays an important role.

The Penrose triangles

This is supplemented by the *laying rules*. They are a bit unusual, but with suitable markings attached to the triangles, they can be easily remembered.

Then it gets a bit technical: Which tiling patterns are possible at all due to the laying rules? Remarkably, the possible constructions can be coded by suitable 0 -1 -sequences (so-called *index sequences*). The main result then states that admissible tiling patterns by Penrose triangles on the one hand and index sequences on the other hand are in a one-to-one relationship and that translational symmetry is never to be expected.

We then turn to the question of how many *different Penrose tilings* there are. Using the description by index sequences, it is possible to show that one can find uncountably many essentially different such tilings, where the difference only becomes apparent when one knows the complete tilings: "Locally", Penrose tilings are indistinguishable from each other.

The book is very detailed. For readers who have successfully completed the basic lectures, it should also be suitable for self-study. At all suitable points, an attempt was made to illustrate the mathematical facts with pictures[2]. And everyone is invited again to "do it by hand" or with computer help to better understand the mathematical background through visualization.

---

[2] The pictures contained here were taken by me or generated with the "Delphi" program.

Part I
Looking over Escher's Shoulder

Symmetries and Fundamental Domains

2

There is no doubt that *Maurits Cornelis Escher* is the main character of the first part of this book. Escher lived from 1898 to 1972. After his studies, he traveled to Italy and Spain several times, and was particularly impressed by the works of Moorish artists in the Alhambra in Granada. At first, the focus of his work was on Mediterranean landscape paintings, later "impossible" pictures and – inspired by the pictures of the Alhambra – tessellations came along.

Escher was a successful artist. A particularly important role was played by the *World Congress of Mathematicians* organized in Amsterdam in 1954. The resulting meetings were very impressive for both sides. The mathematicians could hardly believe that an artist without a mathematical education had, by the power of his intuition, penetrated so deeply into the topic of "symmetry". And Escher was surprised to learn that his constructions, found by trial and error and time-consuming, were part of a theory that was already mathematically complete.

Escher once expressed it very poetically: He had to struggle through the "garden" of mathematics, without knowing that there were already well-prepared paths to his goal.

The present chapter contains some necessary preparations, which are required for the clarification and answering the question "How many different plane symmetry groups are there?".

2.1   What is Symmetry?

It is clear that a circle or a square are "somehow symmetrical". But what does that mean exactly? For this, two questions must be clarified first:

- What is an image?
- What is a movement?

© The Author(s), under exclusive license to Springer Fachmedien Wiesbaden GmbH, part of Springer Nature 2022
E. Behrends, *Tilings of the Plane*, Mathematics Study Resources 2,
https://doi.org/10.1007/978-3-658-38810-2_2

We will only consider *images in the plane*. In the simplest version, it is about a black and white image $\mathbb{B}$, which can be identified with a *subset* $\mathbb{B} \subset \mathbb{R}^2$. If you want to be more precise, you can also take into account the shades of gray, then an image is a mapping ϕ from $\mathbb{R}^2$ to $[0, 1]$, and $\phi(\mathbf{x})$ specifies in which shade of gray $\mathbf{x}$ is colored (somewhere between $\phi(\mathbf{x}) = 0$ for deep black and $\phi(\mathbf{x}) = 1$ for white). You can also take into account colors, which we represent as RGB colors in the number cube $\mathbb{W} := \{0, \ldots, 255\}^3$, where the three numbers stand for the respective red, green and blue components. It is customary to represent them hexadecimally, for example as $\$0B12FF$ (then $0B_{\text{Hex}} = 11$ is the red component, $12_{\text{Hex}} = 18$ the green component, $FF_{\text{Hex}} = 255$ the blue component). $\$000000$ is deep black and $\$FFFFFF$ is white.

Definition 2.1.1
A *picture* is a mapping $\phi : \mathbb{R}^2 \to \{0, \ldots, 255\}^3$. The interpretation: $\mathbf{x}$ is colored according to $\phi(\mathbf{x})$ (RGB-Code).

Such a picture can be rotated, mirrored, moved, etc. We only want to guarantee that mutual distances are preserved:

Definition 2.1.2
(i) Under a *movement* we understand a bijective mapping T on the $\mathbb{R}^2$, which is *distance-preserving*. This means that $\|T\mathbf{x} - T\mathbf{y}\| = \|\mathbf{x} - \mathbf{y}\|$ applies to all $\mathbf{x}, \mathbf{y}$ [1].
(ii) If T is a movement and $\phi : \mathbb{R}^2 \to \{0, \ldots, 255\}^3$ is an image, it can be transformed according to T. $T(\phi)$ should be the image defined by

$$T(\phi)(\mathbf{x}) := \phi\big(T^{-1}(\mathbf{x})\big).$$

(iii) T is called a *symmetry* for ϕ, if $T(\phi) = \phi$.

Remarks
1. The definition of $T(\phi)$ is unusual because T^{-1} appears on the right side. But this guarantees that the image is moved just as T specifies: If T moves one unit to the right, the image moves one unit to the right, etc.
2. You can imagine "symmetry" somewhat simplified by painting the image once on two (infinitely large) transparent foils. Symmetries are then the different ways in which you can superimpose the images.

---

[1] Here $\|\mathbf{x} - \mathbf{y}\|$ denotes the usual (Euclidean) distance between $\mathbf{x}$ and $\mathbf{y}$: If $\mathbf{x}$ or $\mathbf{y}$ has the coordinates x_1, x_2 or y_1, y_2, then $\|\mathbf{x} - \mathbf{y}\| := \sqrt{(x_1 - y_1)^2 + (x_2 - y_2)^2}$.

Examples
1. If ϕ is a constant mapping, the $\mathbb{R}^2$ is thus colored with only one color, then all T are symmetries.
2. The symmetries of a line consist (in addition to the identity) of

   - all translations that are parallel to this line;
   - in addition, the reflection at this line;
   - in addition, all reflections at lines orthogonal to this line;
   - in addition, all combinations of these mappings

3. All rotations around the center are symmetries of a circle. In addition, there are reflections in lines through the center.
4. The identical image is always a symmetry, and sometimes there are no others. (For example, if the image is the letter F. In general, it is helpful to make the symmetries of the different letters clear.)

With S, T it is also obvious that S^{-1} and $S \circ T$ are symmetries of an image, they form a subgroup of the group of bijective maps on $\mathbb{R}^2$: This is the *symmetry group* of the image.

In many cases we will be dealing with images that actually occupy the entire $\mathbb{R}^2$, but of course only a part can be represented. So you have to continue the image "in the same way". Imagine, for example, a section of an infinite chessboard pattern.

2.2   What Movements are There?

Examples of movements are quickly found:

Definition 2.2.1

(i)   Let $\mathbf{a} \in \mathbb{R}^2$ be a vector. With $T_\mathbf{a}$ we denote the *translation* $\mathbf{x} \mapsto \mathbf{a} + \mathbf{x}$ by the vector $\mathbf{a}$.

(ii)   Let $\alpha \in \mathbb{R}$. The *rotation* by the angle α (with the origin as the center of rotation) is described by the mapping

$$R_\alpha \begin{pmatrix} x \\ y \end{pmatrix} = \begin{pmatrix} \cos\alpha & -\sin\alpha \\ \sin\alpha & \cos\alpha \end{pmatrix} \begin{pmatrix} x \\ y \end{pmatrix}.$$

(ii)' If not the origin, but an arbitrary point is the center of rotation, then this results in the *rotation*

$$R_{\alpha,\mathbf{x}_0}(\mathbf{x}) := R_\alpha(\mathbf{x} - \mathbf{x}_0) + \mathbf{x}_0.$$

(iii) Let $\beta \in \mathbb{R}$. With G_β we denote the line that goes through the origin and with the positive direction of the x -axis encloses the angle β:

$G_\beta = \{(t\cos\beta, t\sin\beta)^\top \mid t \in \mathbb{R}\}$. (Here $(a, b)^\top$ stands for the column vector with entries a, b). If you want to reflect an $\mathbf{x}$ on this line, this is done by the mapping

$$S_\beta \begin{pmatrix} x \\ y \end{pmatrix} = \begin{pmatrix} \cos(2\beta) & \sin(2\beta) \\ \sin(2\beta) & -\cos(2\beta) \end{pmatrix} \begin{pmatrix} x \\ y \end{pmatrix}.$$

(That the *reflection* really has this form can be seen by calculating the images of the unit vectors. Or by writing S_β in the form $R_{-\beta} \circ S_0 \circ R_\beta$, where $S_0 = \left(\begin{smallmatrix} 1 & 0 \\ 0 & -1 \end{smallmatrix} \right)$ means reflection at the x -axis. Here one must still remember the formulas $\sin(2\beta) = 2\cos\beta\sin\beta$ and $\cos(2\beta) = \cos^2\beta - \sin^2\beta$.)

(iii)′   Now we want to reflect the line G_β displaced by a vector $\mathbf{g}_0$. This *reflection* can be represented explicitly as $\mathbf{x} \mapsto S_\beta(\mathbf{x} - \mathbf{g}_0) + \mathbf{g}_0$. If it is written as $\mathbf{x} \mapsto \mathbf{g}_0 - S_\beta\mathbf{g}_0 + S_\beta\mathbf{x}$, then $\mathbf{s} := \mathbf{g}_0 - S_\beta\mathbf{g}_0$ is obviously a vector with $S_\beta\mathbf{s} = -\mathbf{s}$.

And vice versa: If $\mathbf{s}$ is given with $S_\beta\mathbf{s} = -\mathbf{s}$, then $\mathbf{x} \mapsto \mathbf{s} + S_\beta\mathbf{x}$ is the reflection at the line displaced by $\mathbf{s}/2$. This follows from the fact that all points $\mathbf{s}/2 + \mathbf{x}$ with $S_\beta\mathbf{x} = \mathbf{x}$, i.e. all points of the displaced line G_β, are fixed under $\mathbf{x} \mapsto \mathbf{s} + S_\beta\mathbf{x}$. We write $S_{\beta,\mathbf{s}}$ for such a reflection[2].

(iii)″   A *glide reflection* $S_{\beta,\mathbf{s},\mathbf{b}}$ is a reflection of the type $S_{\beta,\mathbf{s}}$, combined with a non-trivial translation by the vector $\mathbf{b} \neq \mathbf{0}$, where $\mathbf{b}$ points in the direction of the line being reflected on. Such a $\mathbf{b} \neq \mathbf{0}$ is given, and then $S_{\beta,\mathbf{s},\mathbf{b}} := T_\mathbf{b} \circ S_{\beta,\mathbf{s}}$.

Remark It should be noted that in the definition of translations and rotations we also allowed trivial cases: The identity is a translation by the zero vector and a rotation by a multiple of 2π. However, for glide reflections we require that they are not reflections.

And so one can imagine these movements:

The original and the application of a translation, a rotation and a glide reflection

---

[2] We will always assume that $S_\beta\mathbf{s} = -\mathbf{s}$ is true. It is the reflection at the line perpendicular to $\mathbf{s}/2$.

Proposition 2.2.2
All mappings $T_{\mathbf{a}}, R_{\alpha,\mathbf{x}_0}, S_{\beta,\mathbf{s},\mathbf{b}}$ are movements.

Proof

(i)   It is $\|T_{\mathbf{a}}\mathbf{x} - T_{\mathbf{a}}\mathbf{y}\| = \|(\mathbf{a} + \mathbf{x}) - (\mathbf{a} + \mathbf{y})\| = \|\mathbf{x} - \mathbf{y}\|$. The mapping $T_{\mathbf{a}}$ is certainly bijective, because $T_{-\mathbf{a}}$ is obviously inverse to $T_{\mathbf{a}}$.

(ii)   Let $\mathbf{x} = (x,y)^{\top}$. Then it is

$$\begin{aligned}
\|R_{\alpha}\mathbf{x}\|^2 &= \|\left(x\cos\alpha - y\sin\alpha, x\sin\alpha + y\cos\alpha\right)^{\top}\|^2 \\
&= (x\cos\alpha - y\sin\alpha)^2 + (x\sin\alpha + y\cos\alpha)^2 \\
&= x^2 + y^2 \\
&= \|\mathbf{x}\|^2.
\end{aligned}$$

And thus $\|R_{\alpha}\mathbf{x} - R_{\alpha}\mathbf{y}\| = \|R_{\alpha}(\mathbf{x} - \mathbf{y})\| = \|\mathbf{x} - \mathbf{y}\|$. Note that $R_{-\alpha}$ is inverse to R_{α}.

(ii)'   It is $R_{\alpha,\mathbf{x}_0}\mathbf{x} = R_{\alpha}\mathbf{x} + \mathbf{x}_0 - R_{\alpha}\mathbf{x}_0$. The translation vector $\mathbf{x}_0 - R_{\alpha}\mathbf{x}_0$ cancels out in the difference.

(iii)   That $\|S_{\beta}\mathbf{x}\|^2 = \|\mathbf{x}\|^2$ always holds is calculated directly as in (i), and as in (i) it follows that this mapping is an isometry. S_{β} is inverse to itself.

(iii)' $S_{\beta,\mathbf{s}}$ arises from S_{β} by a translation.

(iii)''   Here only another translation is added.   □

We want to describe movements in a more uniform way. For this we define:

Definition 2.2.3
A real 2×2 -matrix

$$O = \begin{pmatrix} a_{11} & a_{12} \\ a_{21} & a_{22} \end{pmatrix}$$

is called *orthogonal*, if the columns are orthogonal unit vectors, that is, if

$$a_{11}^2 + a_{12}^2 = a_{21}^2 + a_{22}^2 = 1, a_{11}a_{12} + a_{21}a_{22} = 0$$

holds.

Lemma 2.2.4
Let O be orthogonal.

(i)   The line vectors are also normalized and orthogonal, the determinant is 1 or -1.

(ii) $\mathbf{x} \mapsto O\mathbf{x}$ is a movement.

(iii) If $\det O = 1$ and O is not the identity matrix, then there is an $\alpha \in \mathbb{R}$ with $O = R_\alpha$, and conversely, every R_α is orthogonal with determinant 1.

(iv) If $\det O = -1$, then there is a $\beta \in \mathbb{R}$ with $O = S_\beta$, and conversely, every S_β is orthogonal with determinant -1.

(v) The product of orthogonal maps is orthogonal.

(vi) As before, we represent the rotation R_α by $\left(\begin{smallmatrix} \cos\alpha & -\sin\alpha \\ \sin\alpha & \cos\alpha \end{smallmatrix} \right)$ and the reflection S_β by $\left(\begin{smallmatrix} \cos(2\beta) & \sin(2\beta) \\ \sin(2\beta) & -\cos(2\beta) \end{smallmatrix} \right)$. Then we have:

$$R_\alpha \circ R_{\alpha'} = R_{\alpha+\alpha'}, \qquad S_{\beta_1} \circ S_{\beta_2} = R_{2(\beta_1-\beta_2)}.$$
$$R_\alpha \circ S_\beta = S_{\alpha/2+\beta}, \qquad S_\beta \circ R_\alpha = S_{\beta-\alpha/2}.$$

Proof (i) The assumption states that $O^\top O = \mathrm{Id}$. So $\mathbf{x} \mapsto O\mathbf{x}$ is surjective and therefore – since $\mathbb{R}^2$ is finite-dimensional – also injective. $\mathbf{x} \mapsto O^\top \mathbf{x}$ is therefore the inverse map of $\mathbf{x} \mapsto O\mathbf{x}$, and therefore $OO^\top = \mathrm{Id}$ also holds. Consequently, the row vectors are normalized and orthogonal.

It is $1 = \det \mathrm{Id} = \det(OO^\top) = \det O \det O^\top = (\det O)^2$, and that shows that $\det O \in \{-1, 1\}$.

(ii) For any $\mathbf{x}$ one has

$$\|O\mathbf{x}\|^2 = \langle O\mathbf{x}, O\mathbf{x}\rangle = \langle \mathbf{x}, O^\top O\mathbf{x}\rangle = \langle \mathbf{x}, \mathbf{x}\rangle = \|\mathbf{x}\|^2.$$

Because of the linearity of O the claim follows.

(iii) Obviously all R_α are orthogonal with determinant 1 . Conversely: The first column of O must be, because of (ii), a unit vector, we write it as $(\cos\alpha, \sin\alpha)^\top$. Denote the components of the second column with a, b so it follows from the conditions (columns are orthogonal unit vectors, determinant 1), that

$$a^2 + b^2 = 1,$$
$$a\cos\alpha + b\sin\alpha = 0,$$
$$b\cos\alpha - a\sin\alpha = 1.$$

This system of equations has only one solution, namely $a = -\sin\alpha, b = \cos\alpha$.

(iv) The proof is as in (iii): Write the first column of O as $(\cos 2\beta, \sin 2\beta)^\top$ and conclude from the conditions that the second column must be equal to $(\sin 2\beta, -\cos 2\beta)^\top$.

(v) Obviously $O_1 O_1^\top = O_2 O_2^\top = \mathrm{Id}$ follows from $(O_1 O_2)(O_1 O_2)^\top = \mathrm{Id}$.

(vi) This is easily verified using the matrix calculation rules and the addition theorems for sine and cosine.   □

Proposition 2.2.5

If T is a movement, then there is $\mathbf{a} \in \mathbb{R}^2$ and an orthogonal matrix O with $T\mathbf{x} = \mathbf{a} + O\mathbf{x}$ for all $\mathbf{x}$. Here, $\mathbf{a}$ and O are uniquely determined.

Proof That $\mathbf{a}$ and O are uniquely determined is clear, because $\mathbf{a} = T\mathbf{0}$, and $O\mathbf{x} = T\mathbf{x} - \mathbf{a}$.

First, let T be an isometry that fixes the vectors $(0,0)^{\top}, (1,0)^{\top}$ and $(0,1)^{\top}$. Let $(a,b)^{\top}$ be arbitrary and $(c,d)^{\top} := T\big((a,b)^{\top}\big)$. From the isometry condition it follows:

$a^2 + b^2 = c^2 + d^2$, because the distance to $(0,0)^{\top}$ remains unchanged;

$(a-1)^2 + b^2 = (c-1)^2 + d^2$, because the distance to $(1,0)^{\top}$ remains unchanged;

$a^2 + (b-1)^2 = c^2 + (d-1)^2$, because the distance to $(0,1)^{\top}$ remains unchanged.

These are three equations for c, d, which have the only solution $c = a$ and $d = b$. T must therefore be the identity.

Now we start from a T that fixes $(0,0)^{\top}$ and $(0,1)^{\top}$. We write $T((0,1)^{\top})$ as $(c,d)^{\top}$. From the isometry condition it follows that both $c^2 + d^2 = 1$ and $(c-1)^2 + d^2 = 2$ hold. This implies $c = 0$ and $d = \pm 1$. In the case $d = 1$, T must be the identity due to the previous part of the proof. If $d = -1$, we go to $S_0 \circ T$, where S_0 again denotes the reflection in the y-axis. $S_0 \circ T$ then fixes three points as in the first part of the proof, so it is the identity.

In the general case, choose a translation $T_{\mathbf{a}}$ and a rotation R_α so that $T' := R_\alpha \circ T \circ T_{\mathbf{a}}$ fixes the points $(0,0)^{\top}$ and $(0,1)^{\top}$. (First move by $-T\mathbf{0}$ and then rotate $T((1,0)^{\top} - T\mathbf{0})$ – a vector of length 1 – so that it coincides with $(1,0)^{\top}$. Then T' or $S_0 \circ T'$ is the identity, which means $T = T_{-\mathbf{a}} \circ R_{-\alpha}$ or $T = S_0 \circ T_{-\mathbf{a}} \circ R_{-\alpha}$, and that is the composition of an orthogonal transformation with a translation. $\qquad\square$

Proposition 2.2.6

Let T be a movement, $T\mathbf{x} = \mathbf{a} + O\mathbf{x}$. There will occur precisely one of the following cases:

(i) $\mathbf{a} = 0$ and $O = \mathrm{Id}$: Then T is the identical mapping.

(ii) $\mathbf{a} \neq 0$ and $O = \mathrm{Id}$: Then T is the translation $T_{\mathbf{a}}$.

(iii) $O \neq \mathrm{Id}$ and $\det O = 1$: Then T is a rotation. More precisely: O can be written as R_α for a $\alpha \notin 2\pi\mathbb{Z}$, $\mathrm{Id} - R_\alpha$ is invertible, and with $\mathbf{x}_0 := (\mathrm{Id} - R_\alpha)^{-1}\mathbf{a}$ is $T = R_{\alpha,\mathbf{x}_0}$.

(iv) $O \neq \mathrm{Id}$ and $\det O = -1$: Then T is a reflection or a glide reflection. More precisely: We write $\mathbf{a}$ as $\mathbf{a} = \mathbf{s} + \mathbf{b}$, where $O\mathbf{s} = -\mathbf{s}$ and $O\mathbf{b} = \mathbf{b}$. (Set $\mathbf{s} = (\mathbf{a} - O\mathbf{a})/2$, $\mathbf{b} = (\mathbf{a} + O\mathbf{a})/2$. Because of Lemma 2.2.4 (v) O is of the form S_β. So $O^2 = \mathrm{Id}$, and it follows that $\mathbf{s}, \mathbf{b}$ have the claimed properties.) If $\mathbf{b} = 0$, then T is the reflection $S_{\beta,\mathbf{s}}$, and otherwise it is the glide reflection $S_{\beta,\mathbf{s},\mathbf{b}}$.

Proof

(i)   and (ii) are clear.

(iii)   Because of $\det O = 1$ and $O \neq \mathrm{Id}$, $O = R_\alpha$ is true for a suitable $\alpha \in \mathbb{R} \setminus 2\pi\mathbb{Z}$. We want to find an $\mathbf{x}_0$ so that $\mathbf{a} + O\mathbf{x} = R_{\alpha,\mathbf{x}_0}\mathbf{x}$ is true for all $\mathbf{x}$, i. e. $\mathbf{a} = \mathbf{x}_0 - R_\alpha\mathbf{x}_0 = (\mathrm{Id} - R_\alpha)\mathbf{x}_0$. Now $\mathrm{Id} - R_\alpha$ has the determinant $(1 - \cos\alpha)^2 + \sin^2\alpha = 2 - 2\cos\alpha$. This number is different from zero by assumption, and therefore $\mathbf{x}_0$ can be uniquely determined to be $\mathbf{a}$.

(iv)   This is clear from the definitions of $S_{\beta,\mathbf{a}_1}$ and $S_{\beta,\mathbf{a}_1,\mathbf{a}_2}$. $\square$

As a result of the above proposition, it is easy to classify compositions of motions:

- If T_1, T_2 are translations, then $T_1 \circ T_2$ is also a (possibly trivial) translation.
- If T_1, T_2 are rotations, then $T_1 \circ T_2$ is a translation or also a (possibly trivial) rotation.
- If T_1, T_2 are reflections, then $T_1 \circ T_2$ is the identity, a translation, or a rotation.
- If T_1, T_2 are glide reflections, then $T_1 \circ T_2$ is the identity, a translation, or a rotation.
- The product of a translation and a rotation is a rotation.
- The product of a translation and a reflection is a reflection or a glide reflection.
- The product of a translation and a glide reflection is a reflection or a glide reflection.
- The product of a rotation and a reflection is a reflection or a glide reflection.
- The product of a rotation and a glide reflection is a reflection or a glide reflection.
- The product of a reflection and a glide reflection is a translation or a rotation.

(It could be expressed also more detailed: $R_\alpha \circ R_{\alpha'} = R_{\alpha+\alpha'}$, etc.)

2.3   Groups of Movements

Products and inverses of movements are movements, and so is the identity. Consequently, the set of all movements with "product" as inner composition is a subgroup of the bijective mapping on the $\mathbb{R}^2$. We will refer to this subgroup as $\mathcal{G}_0$ below, and we will be interested in subgroups of $\mathcal{G}_0$, motivated by symmetry groups.

We remind you once again of a definition from Sect. 2.1: If ϕ is an image, then the *symmetry group* of ϕ is the set of all symmetries of ϕ, i. e. the set of all movements that leave ϕ invariant (see Definition 2.1.2). This is obviously a subgroup of $\mathcal{G}_0$, which we write as $\mathcal{G}_\phi$. Not all subgroups are symmetry groups:

Proposition 2.3.1

There is a subgroup of $\mathcal{G}_0$ that does not have the form

Proof Let $\mathcal{G}$ be the subgroup of all translations in $\mathcal{G}_0$. Now let an image $\phi : \mathbb{R}^2 \to \{0, \ldots, 255\}^3$ be given such that all translations are symmetries. Then ϕ is necessarily a constant mapping, and therefore there are even more symmetries (all rotations, reflections, glide reflections). $\mathcal{G}$ can therefore not be the symmetry group of ϕ. $\square$

In most cases there is a natural candidate for the symmetry group to an image, and it is not difficult to prove that it is really the symmetry group. Sometimes the situation is not so clear, for illustration we consider a one-dimensional example. First of all, one should realize that the movements[3] are exactly the maps $T_a : x \mapsto x + a$ (translation by a) and $S_b : x \mapsto 2b - x$ (reflection at b). (For the proof it is helpful to prove that a movement that fixes 0 and 1 must be the identity.)

Now we consider the set $M := \bigcup_{q \in \mathbb{Q}} \{q, q + \sqrt{2}, q + \sqrt{3}\}$. We claim that the symmetry group consists of the translations T_a with rational a. These movements are obviously symmetries. It remains to be shown:

- If a translation T_a is a symmetry, then a is rational.
- There is no b for which S_b is a symmetry.

Both statements follow from the fact that the numbers $\sqrt{2}$ and $\sqrt{3}$ are irrational. Consider for example a symmetry T_a. Because of $0 \in M$, $a \in M$. If a is rational, then everything is shown. But it could also be $a = q + \sqrt{2}$ (or $a = q + \sqrt{3}$) for a $q \in \mathbb{Q}$. $\sqrt{2} \in M$ then implies $a + \sqrt{2} = q + 2\sqrt{2} \in M$, i.e. $q + 2\sqrt{2}$ would have to be of the form $r, r + \sqrt{2}$ or $r + \sqrt{3}$ for a rational r. In all cases a contradiction arises to the irrationality of $\sqrt{2}$ and $\sqrt{3}$. For the remaining statements one argues analogously.

From here on, let $\mathcal{G}$ be a subgroup of $\mathcal{G}_0$. With $T, B \in \mathcal{G}$ the conjugated transformation $B \circ T \circ B^{-1}$ is also in $\mathcal{G}$. This has far-reaching consequences:

Proposition 2.3.2

(i)   Let B and T be in $\mathcal{G}$, where B is written as $B : \mathbf{x} \mapsto \mathbf{c} + V\mathbf{x}$ and T is written as $T : \mathbf{x} \mapsto \mathbf{a} + O\mathbf{x}$ with $\mathbf{c}, \mathbf{a} \in \mathbb{R}^2$ and orthogonal V, O (cf. proposition 2.2.5). If one then writes $S := B \circ T \circ B^{-1}$ as $S : \mathbf{x} \mapsto \mathbf{a}' + O'\mathbf{x}$, the following is true:

$$\mathbf{a}' = \mathbf{c} - VOV^\top \mathbf{c} + V\mathbf{a}, \quad O' = VOV^\top$$

(ii) In particular, if $T = T_{\mathbf{a}}$ is a translation (i.e. $O = \mathrm{Id}$ is true), then S is the translation $\mathbf{x} \mapsto V\mathbf{a} + \mathbf{x}$.

(iii) If $B = T_{\mathbf{c}}$ is a translation (so $V = \mathrm{Id}$ applies), then S is the movement $S : \mathbf{x} \mapsto \mathbf{c} - O\mathbf{c} + \mathbf{a} + O\mathbf{x}$.

---

[3]That is, the distance-preserving transformations.

(iv) If $\mathcal{G}$ contains a reflection S_β and a translation $T_\mathbf{a}$, then $\mathcal{G}$ also contains the reflection $\mathbf{x} \mapsto \mathbf{a} - S_\beta\mathbf{a} + S_\beta\mathbf{x}$.

(v)   If $\mathcal{G}$ contains the rotation $R_{\alpha,\mathbf{x}_0} : \mathbf{x} \mapsto \mathbf{x}_0 - R_\alpha\mathbf{x}_0 + R_\alpha\mathbf{x}$ with center of rotation $\mathbf{x}_0$ and the translation $T_\mathbf{a}$, then $\mathcal{G}$ also contains the rotation $R_{\alpha,\mathbf{a}+\mathbf{x}_0}$.

Proof

(i) This is easy to calculate directly:

$$
\begin{aligned}
\mathbf{a}' + O'\mathbf{x} &= B \circ T \circ B^{-1}\mathbf{x} \\
&= B \circ T\left(V^\top x - V^\top \mathbf{c}\right) \\
&= B\left(\mathbf{a} + OV^\top\mathbf{x} - OV^\top\mathbf{c}\right) \\
&= VOV^\top\mathbf{x} + \mathbf{c} - VOV^\top\mathbf{c} + V\mathbf{a},
\end{aligned}
$$

so it follows that $\mathbf{a}' = \mathbf{c} - VOV^\top\mathbf{c} + V\mathbf{a}$ and $O' = VOV^\top$.

(ii) and (iii) follow directly from (i).

(iv) It is $T_\mathbf{a} \circ S_\beta \circ T_{-\mathbf{a}} : \mathbf{x} \mapsto \mathbf{a} - S_\beta\mathbf{a} + S_\beta\mathbf{x}$, where for $\mathbf{s} := \mathbf{a} - S_\beta\mathbf{a}$ it is obviously true that the relationship $S_\beta\mathbf{s} = -\mathbf{s}$ holds. $T_\mathbf{a} \circ S_\beta \circ T_{-\mathbf{a}}$ is therefore a reflection.

(v) Because of (iii),

$$
S : \mathbf{x} \mapsto (\mathbf{a} - R_\alpha\mathbf{a}) + (\mathbf{x}_0 - R_\alpha\mathbf{x}_0 + R_\alpha)\mathbf{x},
$$

and $(\mathbf{a} - R_\alpha\mathbf{a}) + (\mathbf{x}_0 - R_\alpha\mathbf{x}_0)$ can be written as $(\mathbf{a} + \mathbf{x}_0) - R_\alpha(\mathbf{a} + \mathbf{x}_0)$.   □

The following definition will play an important role in the analysis of motion groups:

Definition 2.3.3

(i) Δ is the set of $\mathbf{a} \in \mathbb{R}^2$, for which $T_\mathbf{a}$ belongs to $\mathcal{G}$. Obviously, Δ is an additive subgroup of the $\mathbb{R}^2$.

(ii) Let $\alpha \in \mathbb{R} \setminus 2\pi\mathbb{Z}$ be a rotation angle. Under Δ_α we understand the set of all $\mathbf{x}_0 \in \mathbb{R}^2$, for which $R_{\alpha,\mathbf{x}_0}$ is in $\mathcal{G}$. That is the set of all rotation centers to the α-rotations in $\mathcal{G}$.

Depending on how $\mathcal{G}$ is presented, Δ can be very different: It is possible that Δ consists only of the zero or coincides with the $\mathbb{R}^2$. In the cases that interest us mainly, Δ will be a point lattice and will typically look like the following examples:

Two examples of typical Δ

Proposition 2.3.4

Let $R_\alpha \in \mathcal{G}$ for an $\alpha \in \mathbb{R} \setminus 2\pi\mathbb{Z}$.

(i)   It holds $\Delta_\alpha + \Delta = \Delta_\alpha$ and $R_\alpha(\Delta) = \Delta$.

(ii) It is $(\mathrm{Id} - R_\alpha)\Delta_\alpha = \Delta$ and therefore $\Delta_\alpha = (\mathrm{Id} - R_\alpha)^{-1}\Delta$ applies.

(iii) Concretely this means $(\mathrm{Id} - R_\alpha)^{-1}\mathbf{x} = (\mathrm{Id} - R_{-\alpha})\mathbf{x}/(2 - 2\cos\alpha)$. For some concrete values[4] of α, this means:

- $\alpha = \pi$: It is $(\mathrm{Id} - R_\pi)^{-1}\mathbf{x} = \mathbf{x}/2$:
- $\alpha = 2\pi/3$: It is $(\mathrm{Id} - R_{2\pi/3})^{-1}\mathbf{x} = (\mathbf{x} - R_{-2\pi/3}\mathbf{x})/3$.
- $\alpha = \pi/2$. It is $(\mathrm{Id} - R_{\pi/2})^{-1}\mathbf{x} = (\mathbf{x} - R_{-\pi/2}\mathbf{x})/2$.
- $\alpha = \pi/3$: It is $(\mathrm{Id}\ \ R_{\pi/3})^{-1}\mathbf{x} = \mathbf{x} - R_{-\pi/3}\mathbf{x}$.

Proof

(i)   The statement $\Delta_\alpha + \Delta \subset \Delta_\alpha$ is a rephrasing of proposition 2.3.2 (v), and the relation $\Delta_\alpha \subset \Delta_\alpha + \Delta$ is clear because of $\mathbf{0} \in \Delta$.

Now let $\mathbf{a} \in \Delta$. The movement $R_\alpha \circ T_\mathbf{a} \circ R_{-\alpha}$ belongs to $\mathcal{G}$, it is the translation $T_{R_\alpha\mathbf{a}}$. That is shown by $R_\alpha(\Delta) \subset \Delta$. Analogously one proves $R_{-\alpha}(\Delta) \subset \Delta$, and by applying R_α to this relation it follows that $\Delta \subset R_\alpha(\Delta)$.

(ii) Let $\mathbf{x}_0 \in \Delta_\alpha$, i. e. $R : \mathbf{x} \mapsto \mathbf{x}_0 - R_\alpha\mathbf{x}_0 + R_\alpha\mathbf{x}$ lies in $\mathcal{G}$. Then also $R \circ R_{-\alpha}$ lies in $\mathcal{G}$, and $R \circ R_{-\alpha} = T_{\mathbf{x}_0 - R_\alpha\mathbf{x}_0}$, i. e. $(\mathrm{Id} - R_\alpha)\mathbf{x}_0 \in \Delta$. That is shown by $(\mathrm{Id} - R_\alpha)\Delta_\alpha \subset \Delta$.

Conversely, if $\mathbf{a} \in \Delta$ is given, then $T_\mathbf{a} \circ R_\alpha : \mathbf{x} \mapsto \mathbf{a} + R_\alpha\mathbf{x}$ lies in $\mathcal{G}$. Write $\mathbf{a}$ as $\mathbf{a} = \mathbf{x}_0 - R_\alpha\mathbf{x}_0$: That is possible, since $\mathrm{Id} - R_\alpha$ is bijective (cf. proposition 2.2.6 (iii)). Then $\mathbf{x}_0 \in \Delta_\alpha$, thus $\mathbf{a} \in (\mathrm{Id} - R_\alpha)\Delta_\alpha$.

---

[4]Why these values are considered will become clear in Section 3.5 .

(iii) From the matrix representation of R_α it immediately follows that $R_\alpha + R_{-\alpha} = (2\cos\alpha)\,\mathrm{Id}$. From that we conclude

$$(2 - 2\cos\alpha)\mathrm{Id} = (\mathrm{Id} - R_\alpha)(\mathrm{Id} - R_{-\alpha}),$$

and this equation immediately implies the claim.   □

And what happens if $\mathcal{G}$ contains reflections or glide reflections?

Proposition 2.3.5

Let the reflection S_β or the glide reflection $G : \mathbf{x} \mapsto \mathbf{b} + S_\beta\mathbf{x}$ be in $\mathcal{G}$. In the second case one has $S_\beta\mathbf{b} = \mathbf{b}$. Then $S_\beta(\Delta) = \Delta$.

Proof It is enough to focus on glide reflections and possibly set $b = 0$. We apply proposition 2.3.2: With $G \circ T_{\mathbf{c}} \circ G^{-1}$ also

$$G \circ T_{\mathbf{c}} \circ G^{-1}\mathbf{x} = S_\beta\mathbf{c} + \mathbf{x}.$$

is in the group, and $G \circ T_{\mathbf{c}} \circ G^{-1} = T_{S_\beta\mathbf{c}}$ Therefore, $S_\beta(\Delta) \subset \Delta$. This proves S_β^2, and since $S_\beta(\Delta) = \Delta$ is the identity, even S_β applies.   □

Proposition 2.3.6

Let S_β be a reflection and $b \in \mathbb{R}^2$, and B denotes the movement $B : x \mapsto b + S_\beta x$

(i)   Any two of the statements $B \in \mathcal{G}, b \in \Delta$ and $S_\beta \in \mathcal{G}$ imply the third.

(ii)   Suppose that in addition $S_\beta\mathbf{b} = \mathbf{b}$ as well as $\mathbf{b} \neq \mathbf{0}$, i. e. B is a glide reflection. B lies in $\mathcal{G}$. Then $2\mathbf{b} \in \Delta$. There are two cases possible:

Case 1: One even has $\mathbf{b} \in \Delta$. Then $S_\beta \in \mathcal{G}$, and $B = T_{\mathbf{b}} \circ S_\beta$. Such a glide reflection is called an *improper glide reflection*.

Case 2: $\mathbf{b} \notin \Delta$.   Then B in $\mathcal{G}$ cannot be decomposed into a translation and a reflection. We will call B a *proper glide reflection* in this case[5].

Proof

(i)   The first part follows from $B \circ S_\beta = T_\mathbf{b}$ and $B = T_b \circ S_\beta$.

(ii) This follows from $B^2 = T_{2\mathbf{b}}$.   □

---

[5] Note: Whether a glide reflection is proper or not depends on $\mathcal{G}$.

Finally, we investigate the case where rotations and reflections belong to $\mathcal{G}$ at the same time:

Proposition 2.3.7

Suppose that $\mathcal{G}$ contains a reflection $S = S_{\beta,s} : \mathbf{x} \mapsto \mathbf{s} + S_\beta \mathbf{x}$ and an α-rotation with center of rotation $\mathbf{x}_0$.

(i)   Then $\mathcal{G}$ also contains the $-\alpha$-rotation with center of rotation $S\mathbf{x}_0$.

(ii)   If α is of the special form $2\pi/n$ for some $n \in \mathbb{N}$, then $\mathcal{G}$ also contains the α-rotation with center of rotation $S\mathbf{x}_0$.

Proof (i) We observe that $S_\beta \circ R_\alpha \circ S_\beta = R_{-\alpha}$ (cf. Lemma 2.2.4 (vi)), in particular, $S_\beta \circ R_\alpha \mathbf{s} = -S_\beta \circ R_\alpha \circ S_\beta \mathbf{s} = -R_{-\alpha}\mathbf{s}$, because $S_\beta \mathbf{s} = -\mathbf{s}$.

Set $\mathbf{a} = \mathbf{x}_0 - R_\alpha \mathbf{x}_0$, the disputed rotation thus has the form $R : \mathbf{x} \mapsto \mathbf{a} + R_\alpha \mathbf{x}$. Then

$$S \circ R \circ S : \mathbf{x} \mapsto \mathbf{s} + S_\beta \mathbf{a} + S_\beta R_\alpha \mathbf{s} + R_{-\alpha}\mathbf{x}.$$

The claim then amounts to the equation

$$\mathbf{s} + S_\beta \mathbf{a} + S_\beta R_\alpha \mathbf{s} = (\mathbf{s} + S_\beta \mathbf{x}_0) - R_{-\alpha}(\mathbf{s} + S_\beta \mathbf{x}_0),$$

which is easily verified using the preparations, the identity $S_\beta \circ R_\alpha = R_{-\alpha} \circ S_\beta$ and the equation $S_\beta \mathbf{s} = -\mathbf{s}$.

(ii) Rotation centers for α'-rotations are also rotation centers for $k\alpha'$-rotations. Note that here $\alpha = -(n-1)\alpha$ modulo 2π.   □

2.4   Discontinuous Groups and Fundamental Domains

Suppose one knows that an image has a reflection in a line G as symmetry. Then one can reconstruct it if one only knows it on one side of G. The same applies if a rotation by 180 degrees is a symmetry and the point of rotation is on G.

A fundamental domain (*red*) for a reflection or for a rotation

Even less information is needed if there is a rotation by a 45 -degree angle as symmetry, because then only the area between two half-lines with inner angle 45 degrees is sufficient to learn everything about the image.

A fundamental domain (*red*) for a 45 -degree rotation

As a final preparatory example, we assume that there are translation symmetries for two translations in linearly independent directions. Then it is enough to know the image for the points in a parallelogram.

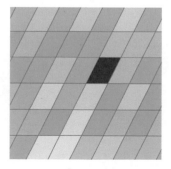

A fundamental domain (*red*) for a group generated by 2 translations

In general, the following problem arises:

Let $\mathcal{G}$ be the symmetry group of an image; $\mathcal{G}$ is therefore a subgroup of the movements, and each $T \in \mathcal{G}$ carries the image into itself. Find then an "as small as possible" subset $F \subset \mathbb{R}^2$ so that

$$\mathbb{R}^2 = \bigcup_{T \in \mathcal{G}} T(F).$$

This would mean: If one knows the F section of the image, one can reconstruct the entire image.

In the preceding examples, we found such F:

- Reflection at a line G and 180 -degree rotation with center of rotation at G: The points on one side of G (including G);
- 45 -degree rotation: The points in a 45 -degree angular space;
- Two translations: The points in a parallelogram.

The examples already show that it is not possible, or only very artificially, to achieve in condition $\mathbb{R}^2 = \bigcup_{T \in \mathcal{G}} T(F)$ that it is a *disjoint* union, that is, that F was really chosen to be smallest. (In the last example, for example, one could remove some pieces from the sides of the parallelogram.) It can also happen that – as in the case of the circle – *all* rotations are symmetries, and then one could choose F to be a half-line[6]. That is not desirable, because the restriction of the image to F should represent a characteristic section of the image. For this reason, one defines:

Definition 2.4.1
A group of movements $\mathcal{G}$ is called *discontinuous* (or *discrete*), if there is an $\mathbf{x_0}$ and a $\varepsilon > 0$ with the following property: If $T \in \mathcal{G}$ and $T \neq \mathrm{Id}$, then $\|\mathbf{x_0} - T\mathbf{x_0}\| \geq \varepsilon$.

The symmetry groups of a circle or a line are thus not discontinuous, but the images that interest us here have discontinuous symmetry groups. Then we also have a chance to find areas F that are "very small" and represent typical aspects of the image:

---

[6]Even more extreme would be if all movements occurred as symmetries, then even a one-pointed F would be possible.

Definition 2.4.2

Let $\mathcal{G}$ be a symmetry group. A subset $F \subset \mathbb{R}^2$ is called a *fundamental domain* for $\mathcal{G}$ if the following conditions are satisfied:

- F is closed and F is the closure of the interior of F. Usually, to avoid pathologies, one also assumes that F is connected or even simply connected.
- $\mathbb{R}^2 = \bigcup_{T \in \mathcal{G}} T(F)$.
- If $S, T \in \mathcal{G}$ and $S \neq T$ intersect, then $S(F)$ and $T(F)$ intersect only in the boundary of $S(F)$.

If one has found a fundamental domain F for a group of movements, one can be creative and generate own images. One just has to design F in an artistically interesting way and then apply all motions of the group to it. The resulting image has (at least) the given group as symmetry group.

The Discontinuous Symmetry Groups of the Plane

3

First, we clarify what we mean in this book by two movement groups being "essentially equal". Then the discrete groups are characterized in terms of the richness of the contained translation subgroup:

- Case 1: There are no non-trivial translations (Leonardo's Theorem, Sect. 3.2).
- Case 2: There is a one-dimensional set of translations (the seven Frieze groups, Sect. 3.4).
- Case 3: There is a two-dimensional set of translations (the seventeen plane crystal groups, Sect. 3.5).

3.1 How Many Different Groups of Movements are There?

Mathematicians try to focus on the essentials in all theories. Here the question is: How many essentially different symmetry groups are there? For this one has to decide what it means that two symmetry groups are "essentially equal". We discuss several approaches.

Attempt 1: identical subgroups of the group of movements $\mathcal{G}_1$ and $\mathcal{G}_2$ are subgroups of the group of movements. They should only be called "equal" if they are identical. That's not very meaningful. For example, the symmetry groups of non-parallel lines would be different, but it makes sense to identify them if you want to work out the essentials.

Attempt 2: Group isomorphism As another extreme, one could consider two symmetry groups $\mathcal{G}_1$ and $\mathcal{G}_2$ as "equal" if they are isomorphic as groups, that is, if there is a bijection Φ from $\mathcal{G}_1$ to $\mathcal{G}_2$ with the property $\Phi(S \circ T) = \Phi(S) \circ \Phi(T)$ (all S, T).

© The Author(s), under exclusive license to Springer Fachmedien Wiesbaden GmbH, part of Springer Nature 2022
E. Behrends, *Tilings of the Plane*, Mathematics Study Resources 2,
https://doi.org/10.1007/978-3-658-38810-2_3

Then one would have to identify reflections with rotations by 180°, and the subgroup generated by a translation would be equivalent to the one generated by a glide reflection (both are infinite and cyclic, thus isomorphic). But we experience these movements as different.

Attempt 3: Conjugate subgroups It has proven itself in group theory to consider two elements g_1, g_2 of a group $(G, \circ)$ as "essentially equal" if they are *conjugate* to each other, that is, if there is an h so that $g_2 = h \circ g_1 \circ h^{-1}$. And two subgroups U_1, U_2 of G are called conjugate if $U_2 = g \circ U_1 \circ g^{-1} := \{g \circ u \circ g^{-1} \mid u \in U_1\}$. In such a case, U_1, U_2 does not differ much, since in particular $u \mapsto g \circ u \circ g^{-1}$ is a group isomorphism from U_1 to U_2.

Therefore, we provisionally define:

Definition 3.1.1

(i) Two movements S, T are called *conjugate*, if there is a movement B so that $S = B \circ T \circ B^{-1}$. This is obviously an equivalence relation.

(ii) $\mathcal{G}_1$ and $\mathcal{G}_2$ are subgroups of the group of movements. They are to be called *conjugate*, if there is a movement B so that

$$\mathcal{G}_1 = B \circ \mathcal{G}_2 \circ B^{-1} := \{B \circ T \circ B^{-1} \mid T \in \mathcal{G}_2\}.$$

Which properties remain when conjugating?

Lemma 3.1.2

(i)   Suppose that the movements S, T are conjugated.
   a) S is exactly the identity if T is the identity.
   b) S is exactly a translation if T is a translation. The translation vectors then have the same length.
   c) S is exactly a rotation if T is a rotation.
   d) S is a reflection if and only if T is a reflection.
   e) S is a glide reflection if and only if T is a glide reflection.

(ii)   Two translations are conjugate if and only if the translation vectors have the same length.

(iii) Two rotations $R_{\alpha, x_0}, R_{\alpha', x_0'}$ are conjugate if and only if $\alpha = \alpha'$ or $\alpha = -\alpha'$.

(iv)   Two reflections are conjugate.

(v)   Two glide reflections are conjugate if and only if the translation vectors have the same length.

(vi)   Pairwise *not* conjugate are the identity, a translation, a rotation, a reflection, and a glide reflection.

Proof (i) Let S and T be movements: $Sx = \mathbf{a}' + O'\mathbf{x}$, $Tx = \mathbf{a} + O\mathbf{x}$. To conjugate, we use any movement B, that is, a map $B : \mathbf{x} \mapsto \mathbf{c} + V\mathbf{x}$ with an orthogonal V (the inverse is given by $B^{-1}x = V^\top x - V^\top \mathbf{c}$). If $S = B \circ T \circ B^{-1}$, then:

$$\mathbf{a}' + O'\mathbf{x} = B \circ T \circ B^{-1}\mathbf{x}$$
$$= B \circ T(V^\top x - V^\top \mathbf{c})$$
$$= B(\mathbf{a} + OV^\top \mathbf{x} - OV^\top \mathbf{c})$$
$$= VOV^\top \mathbf{x} + \mathbf{c} - VOV^\top \mathbf{c} + V\mathbf{a},$$

so it follows that $O' = VOV^\top$ and $\mathbf{a}' = \mathbf{c} - VOV^\top \mathbf{c} + V\mathbf{a}$.

Now we prove the statements a) to e). Since "equivalence" is a symmetrical relation, only one direction has to be shown in each case.

a) If $T = \mathrm{Id}$, then O is the identity matrix and $\mathbf{a} = \mathbf{0}$. Then $O' = \mathrm{Id}$ and $\mathbf{a}' = \mathbf{0}$ are also true.

b) If T is a non-trivial translation, then $O = \mathrm{Id}$ and $\mathbf{a} \neq \mathbf{0}$. It follows that $O' = \mathrm{Id}$ and $\mathbf{a}' = V\mathbf{a} \neq \mathbf{0}$, i.e. S is also a translation. Note that V is an isometry and therefore $\|\mathbf{a}'\| = \|V\mathbf{a}\| = \|\mathbf{a}\|$.

c) If T is a non-trivial rotation, then $O = R_\alpha \neq \mathrm{Id}$. We distinguish two cases. First, $\det V = 1$ could be, then V is of the form $R_{\alpha'}$. It is therefore due to the formulas in Lemma 2.2.4 (vi) $O' = R_{\alpha'} R_\alpha R_{-\alpha'} = R_\alpha$, i. e. also S is a rotation. In the case $\det V = -1$, $V = S_\beta$, and then, again because of Lemma 2.2.4 (vi),

$$VR_\alpha V^{-1} = VR_\alpha V$$
$$= S_\beta R_\alpha S_\beta$$
$$= S_\beta S_{\alpha/2+\beta}$$
$$= R_{2(\beta-(\alpha/2+\beta))}$$
$$= R_{-\alpha}.$$

d, e) T is a reflection or glide reflection, i. e., it is $\det O = -1$. According to the determinant product theorem, $\det VOV^{-1} = -1$ is also true, i. e., S is also a reflection or glide reflection.

Now let T be a reflection, i. e. $O\mathbf{a} = -\mathbf{a}$. We claim that $O'\mathbf{a}' = -\mathbf{a}'$ also holds. Indeed, (using $O\mathbf{a} = -\mathbf{a}$ and $O^2 = \mathrm{Id}$, since it is a reflection)

$$O'\mathbf{a}' = VOV^\top \left(\mathbf{c} - VOV^\top \mathbf{c} + V\mathbf{a}\right)$$
$$= VOV^{-1}\mathbf{c} - \mathbf{c} - V\mathbf{a}$$
$$= -\mathbf{a}'.$$

(ii) If $\mathbf{a} \neq \mathbf{0}$ and $\mathbf{a}' \neq 0$ have the same length, then one only has to choose V so that $\mathbf{a}' = V\mathbf{a}$. This is always possible.

(iii) Let $Tx = \mathbf{a} + R_\alpha \mathbf{x}$ and $Sx = \mathbf{a}' + R_\alpha \mathbf{x}$. We conjugate T with a translation $T_\mathbf{c}$ with an unknown $\mathbf{c}$. The result: $\mathbf{c} - R_\alpha \mathbf{c} + \mathbf{a} + R_\alpha \mathbf{x}$. So that it is equal to S,

$\mathbf{a}' = \mathbf{c} - R_\alpha \mathbf{c} + \mathbf{a}$ must be. But this can be achieved: Set $\mathbf{c} = (\mathrm{Id} - R_\alpha)(\mathbf{a}' - \mathbf{a})$. (Note that $\mathrm{Id} - R_\alpha$ is invertible; see proposition 2.2.6 (iii).)

You proceed in an entirely analogous way when $S\mathbf{x} = \mathbf{a}' + R_{-\alpha}\mathbf{x}$. There you conjugate T with $B : \mathbf{x} \mapsto \mathbf{c} + S_\beta \mathbf{x}$, where S_β is any reflection; $\mathbf{c}$ is still free. It is then (because of $S_\beta R_\alpha S_\beta = R_{-\alpha}$, see Lemma 2.2.4 (iv))

$$B \circ S \circ B^{-1}\mathbf{x} = \mathbf{c} + S_\beta \mathbf{a} - R_{-\alpha}\mathbf{c} + R_{-\alpha}\mathbf{x}.$$

This time you have to choose $\mathbf{c}$ so that $\mathbf{c} - R_{-\alpha}\mathbf{c} + S_\beta \mathbf{a} = \mathbf{a}'$. That is possible because $\mathrm{Id} - R_{-\alpha}$ is invertible.

That $\alpha' = \pm\alpha$ must be the case for conjugated rotations, we have already seen above in the proof of (i)c.

(iv) S is an arbitrary reflection. We show that it is conjugated to $S_{0,0}$. That is enough because "is conjugated to" is an equivalence relation.

S has the form $\mathbf{x} \mapsto \mathbf{s} + S_\beta \mathbf{x}$, where $S_\beta \mathbf{s} = -\mathbf{s}$. If we conjugate the translation $T_\mathbf{c}$ with a still free $\mathbf{c}$, we get the mapping $\mathbf{x} \mapsto \mathbf{s} + \mathbf{c} - S_\beta \mathbf{c} + S_\beta \mathbf{x}$ due to the above calculation rules. The special choice $\mathbf{c} := -\mathbf{s}/2$ leads to $\mathbf{x} \mapsto S_\beta \mathbf{x}$. If we still conjugate S_β with a rotation R_α, we get the mapping $S_{\alpha+\beta}$ according to Lemma 2.2.4 (vi), the choice $\alpha = -\beta$ therefore leads to $S_0 = S_{0,0}$.

(v) $S = S_{\beta,\mathbf{s},\mathbf{b}}$ is a glide reflection, that is, $S_\beta \mathbf{s} = -\mathbf{s}$ and $S_\beta \mathbf{b} = \mathbf{b}$. As with reflections, we can conjugate with a translation and thus achieve $\mathbf{s} = \mathbf{0}$. $S_{\beta,0,\mathbf{b}}$ is conjugated further. We conjugate with $R_{-\beta}$, and since $S_\beta \mathbf{b} = \mathbf{b}$ holds, it follows that $S_0(R_{-\beta}\mathbf{b}) = R_{-\beta}\mathbf{b}$:

$$S_0 R_{-\beta}\mathbf{b} = S_{\beta/2}\mathbf{b},$$
$$R_{-\beta}\mathbf{b} = R_{-\beta}S_\beta \mathbf{b} = S_{-\beta/2+\beta}\mathbf{b} = S_{\beta/2}\mathbf{b}.$$

The vector $\mathbf{b}' := R_{-\beta}\mathbf{b}$ therefore points in the direction of the x-axis. If we conjugate with $S_{\beta,0,\mathbf{b}}$ with $S_{0,0}$, we get $S_{\beta,0,-\mathbf{b}}$, and another conjugation with $R_{-\beta}$ leads to $S_{0,0,-\mathbf{b}'}$.

Together: $S_{\beta,\mathbf{s},\mathbf{b}}$ is conjugate to $S_{0,0,\pm\mathbf{b}'}$, where $\pm\mathbf{b}'$ vectors are in the direction of the x axis, which have the same length as $\mathbf{b}$. The claim follows.

(vi) This follows from the specific transformation formulas that were shown at the beginning of the proof. By conjugation, O is transformed into the transformation matrix $V \circ O \circ V^{-1}$. Therefore, for example, a translation cannot be conjugated to a reflection, etc.   □

The preceding results can be summarized as follows:

Each movement is conjugate to one of the transformations listed in the following catalog ($\mathbf{e}$ denotes the vector $(1,0)^\top$, that is, the unit vector in the direction of the positive x axis):

$$\{\mathrm{Id}\} \cup \{T_{t\mathbf{e}} \mid t > 0\} \cup \{R_\alpha \mid 0 < \alpha \le \pi\} \cup \{S_{0,0}\} \cup \{S_{0,0,t\mathbf{e}} \mid t > 0\}.$$

At first glance, this approach looks very promising. If several movements are involved, however, too many situations are still counted as different, which should actually be identified. Let's take two translations $T_{\mathbf{a}_1}$ and $T_{\mathbf{a}_2}$. If they are conjugated with the same movement, they will be translations with translation vectors $V\mathbf{a}_1$ and $V\mathbf{a}_2$. From orthogonal translation vectors, orthogonal directions are also obtained after conjugation. It would make more sense if only the information "two translation directions" remained.

Attempt 4: the "right" definition After the preceding attempts to find the right definition of "equal" for transformation groups, we now give the version that is suitable for our purposes:

Definition 3.1.3

$\mathcal{G}_1$ and $\mathcal{G}_2$ are subgroups of the group of movements. They are called *equivalent* if there is a group isomorphism Φ from $\mathcal{G}_1$ to $\mathcal{G}_2$ such that for $T \in \mathcal{G}_1$ it holds:

   T is a translation exactly when $\Phi(T)$ is a translation.
   T is a rotation exactly when $\Phi(T)$ is a rotation.
   T is a reflection exactly when $\Phi(T)$ is a reflection.
   T is a glide reflection exactly when $\Phi(T)$ is a glide reflection.

Because of the results assembled in Lemma 3.1.2, it is clear that conjugate subgroups are also equivalent. The converse does not have to hold, however: For any two groups that are generated by two translations with linearly independent vectors, the groups are equivalent; but they are conjugate only when the translation vectors of one group can be transformed into the translation vectors of the other group by an orthogonal transformation.

As our first application, we show for illustration:

Proposition 3.1.4

Let $\mathcal{G}$ be a non-trivial cyclic group of movements[1]. Then, up to equivalence, exactly one of the following cases occurs:

(i) $\mathcal{G}$ has two elements. Then $\mathcal{G}$ is generated by a reflection or a rotation by the angle π.
(ii) $\mathcal{G}$ is finite with more than two elements. Then there is a rotation R by the angle $2\pi/n$, so that $\mathcal{G} = \{\mathrm{Id}, R, R^2, \ldots, R^{n-1}\}$.
(iii) $\mathcal{G}$ is infinite, and there is a translation T with $\mathcal{G} = \{\mathrm{Id}, T, T^2, \ldots\}$.

---

[1] $\mathcal{G}$ is thus generated by a single element.

(iv) $\mathcal{G}$ is infinite, and there is a glide reflection G with $\mathcal{G} = \{\mathrm{Id}, G, G^2, \ldots\}$.

Proof First, let $\mathcal{G}$ be finite. The only motions of finite order are rotations by an angle of the form $2\pi/n$ (order n) and reflections (order 2). So the first two statements follow.

And if $\mathcal{G}$ is infinite, then only translations and glide reflections are possible, because an infinite cyclic group generated by a rotation cannot be discontinuous. □

It is easy to see that these groups can also all occur as symmetry groups of concrete images:

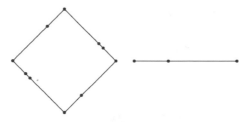

Symmetry groups: Id and a rotation; Id and a reflection

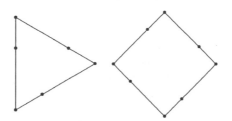

Symmetry groups: 3- and 4-fold rotations

FFFFFFFFFFF

Symmetry groups: generated by translations (detail)

Symmetry groups: generated by glide reflections (detail)

3.2   Finite Groups of Movements

We assume that $\mathcal{G}$ is a finite group of movements that does not consist only of the identity. What possibilities are there? It is obvious that $\mathcal{G}$ cannot contain translations and glide reflections, because they would create infinite subgroups. But we can say more:

Lemma 3.2.1
Suppose that $\mathcal{G}$ is finite.

(i)   If $\mathcal{G}$ contains the rotations $R_{\alpha,\mathbf{x}_0}$, $R_{\alpha',\mathbf{x}_0'}$ with $\alpha, \alpha' \notin 2\pi\mathbb{Z}$, then $\mathbf{x}_0 = \mathbf{x}_0'$.
(ii) If $\mathcal{G}$ contains the reflections $S_{\beta,\mathbf{s}}$ and $S_{\beta,\mathbf{s}'}$ (i.e. it is reflected in parallel lines), then $\mathbf{s} = \mathbf{s}'$.
(iii) If $\mathcal{G}$ contains a reflection and a rotation, then the center of rotation lies on the mirror axis.

Proof (i) Let $S := R_{\alpha,\mathbf{x}_0}$ and $T := R_{\alpha',\mathbf{x}_0'}$ be given, w.l.o.g. let $\mathbf{x}_0' = \mathbf{0}$. The second rotation is therefore $R_{\alpha'}$. Then

$$S \circ T \circ S^{-1} \circ T^{-1}\mathbf{x} = \mathbf{x} + (\mathrm{Id} - R_{\alpha'}) \circ (\mathrm{Id} - R_{\alpha})\mathbf{x}_0.$$

Since $\mathrm{Id} - R_\alpha$ and $\mathrm{Id} - R_{\alpha'}$ are invertible, in the case of $\mathbf{x}_0 \neq \mathbf{0}$ it would be a non-trivial translation in $\mathcal{G}$.

(ii) We have $S_{\beta,\mathbf{s}'} \circ S_{\beta,\mathbf{s}}\mathbf{x} = (\mathbf{s}' - \mathbf{s}) + \mathbf{x}$, and since there are no non-trivial translations, $\mathbf{a}' = \mathbf{a}$ must hold.

(iii) Let T be a rotation. W.l.o.g. let $\mathbf{0}$ be the center of rotation, i. e. it is $T = R_\alpha$, where $\alpha \notin 2\pi\mathbb{Z}$. Further, let $S := S_{\beta,\mathbf{s}}$ be a reflection, so $S_\beta\mathbf{s} = -\mathbf{s}$. It is to be shown that $\mathbf{s} = \mathbf{0}$ holds. For this we calculate $T \circ S \circ T^{-1} \circ S \circ T^{-2}$, again Lemma 2.2.4 (vi) is important:

$$
\begin{aligned}
T \circ S \circ T^{-1} \circ S \circ T^{-2}\mathbf{x} &= T \circ S \circ T^{-1} \circ S\big(R_{-2\alpha}\mathbf{x}\big) \\
&= T \circ S \circ T^{-1}\big(\mathbf{s} + S_\beta R_{-2\alpha}\mathbf{x}\big) \\
&= T \circ S \circ T^{-1}\big(\mathbf{s} + S_{\beta+\alpha}\mathbf{x}\big) \\
&= T \circ S\big(R_{-\alpha}\mathbf{s} + R_{-\alpha}S_{\beta+\alpha}\mathbf{x}\big) \\
&= T \circ S\big(R_{-\alpha}\mathbf{s} + S_{\beta+\alpha/2}\mathbf{x}\big) \\
&= T\big(\mathbf{s} + S_\beta R_{-\alpha}\mathbf{s} + S_\beta S_{\beta+\alpha/2}\mathbf{x}\big) \\
&= T\big(\mathbf{s} + S_\beta R_{-\alpha}\mathbf{s} + R_{-\alpha}\mathbf{x}\big) \\
&= R_\alpha\mathbf{s} + R_\alpha S_\beta R_{-\alpha}\mathbf{s} + R_\alpha R_{-\alpha}\mathbf{x} \\
&= R_\alpha\mathbf{s} + S_{\alpha+\beta}\mathbf{s} + \mathbf{x}
\end{aligned}
$$

Now $S_{\alpha+\beta} = R_{2\alpha}S_\beta$, so $S_{\alpha+\beta}\mathbf{s} = -R_{2\alpha}\mathbf{s}$, i. e. the map in question is the translation by the vector $R_\alpha(\mathrm{Id} - R_\alpha)\mathbf{s}$. If $\mathbf{s} \neq \mathbf{0}$ were so, this vector would be different from zero. (Since $\mathrm{Id} - R_\alpha$ is invertible.)

Here is another alternative proof: The mirror image of the center of rotation is, according to proposition 2.3.7, a center of rotation to a rotation that belongs to the group. Because of part (i) of this proposition, $\mathbf{x}_0$ must agree with its mirror image, so this vector lies on the mirror axis.

$\square$

Definition 3.2.2

(i) For $n \in \mathbb{N}$ let C_n be the cyclic group of order n; one can realize C_n as a group of movements by the rotations $R_{2\pi k/n}$, $k = 0$, $n - 1$. It is also the symmetry group of a regular n-gon Δ_n, which has been completed to a figure Δ_n' in such a way that the rotational symmetries are retained but the reflections are removed. (Compare the following pictures.)

(ii) The *dihedral* group D_n is the symmetry group of a regular n-gon.
   - D_1 is the symmetry group consisting of a reflection, for example the symmetry group of the "dotted" interval Δ_0 in the next image.
   - D_2 is the symmetry group of an interval, that is, it consists of a reflection and a rotation.
   - D_n contains the identity, rotations by $2\pi k/n$ ($k = 1,\ldots,n-1$) and n reflections.

Here are some pictures:

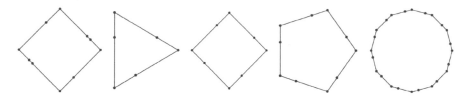

Examples of the symmetry groups $C_2, C_3, C_4, C_5, C_{12}$

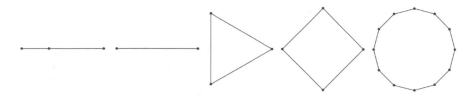

Examples of the symmetry groups $D_1, D_2, D_3, D_4, D_{12}$

The main result of this section is the

> **Theorem 3.2.3 (Leonardo's Theorem)**
> Any finite group of movements is equivalent to a symmetry group C_n or a symmetry group D_n.

Proof Let $\mathcal{G}$ be a non-trivial finite group of movements.

Case 1: $\mathcal{G}$ contains only rotations in addition to the identity

Because of Lemma 3.2.1 (i), the center of rotation is identical for all rotations. W.l.o.g. let it be the origin, so it is about rotations of the type R_α. Suppose that as α-values the numbers $\alpha_1, \ldots, \alpha_k \in \left]0, 2\pi\right[$ occur. α be the smallest of these values. Necessarily (because R_α has finite order) is α of the form $2\pi/n$.

We claim that all $\alpha_1, \ldots, \alpha_k$ are multiples of α. To do this, we consider the numbers $0, \alpha, 2\alpha, \ldots, (n-1)\alpha, 1$. If a α_k *did not* occur in this list, it would be between $l\alpha$ and $(l+1)\alpha$ for a suitable l, but then $R_{(l+1)\alpha}R_{\alpha_k}^{-1} = R_{(l+1)\alpha - \alpha_k}$ would be a rotation with a smaller angle of rotation than α in contradiction to the choice of α. Together this means: $\mathcal{G}$ is the cyclic group C_n and therefore equivalent to the symmetry group of Δ'_n (cf. the definition above).

Case 2: $\mathcal{G}$ contains only one reflection in addition to the identity. Then $\mathcal{G}$ is cyclic with order 2 and equivalent to the D_1.

Case 3: $\mathcal{G}$ contains at least two reflections in addition to the identity. We choose two different reflections S_1 and S_2. The mirror axes cannot be both different and parallel at the same time due to Lemma 3.2.1 (ii). They therefore intersect somewhere—w.l.o.g. at the origin—and have different slopes β_1, β_2, for example $0 \leq \beta_1 < \beta_2 < \pi$. But it is $S_{\beta_1} \circ S_{\beta_2} = R_{2(\beta_1 - \beta_2)}$ (Lemma 2.2.4 (iv)), i. e. $\mathcal{G}$ also contains a non-trivial rotation[2]. All rotations in the subgroup of rotations have $\mathbf{0}$ as the center of rotation. After the first part of the proof (Case 1) there is $\alpha = 2\pi/n$, so that the rotations in $\mathcal{G}$ consist exactly of the $R_{k\alpha}$ ($k = 1, \ldots, n-1$).

We claim that $\mathcal{G}$ is equivalent to the group D_n. We place Δ_n as in the pictures, the corners are the n-th roots of unity.

Now let $S \in \mathcal{G}$ be an arbitrary reflection, for example $\mathbf{x} \mapsto \mathbf{a} + S_\beta \mathbf{x}$. It is necessary that $\mathbf{a} = \mathbf{0}$: If $\mathbf{a} \neq \mathbf{0}$, then $S_{\beta_1} \circ S$ would be the rotation $\mathbf{x} \mapsto S_\beta \mathbf{a} + R_{2(\beta_1 - \beta)}$ (Lemma 2.2.4(iv)). The center of rotation would be different from zero. But there is also a rotation with center of rotation $\mathbf{0}$ in $\mathcal{G}$, and because of Lemma 3.2.1(i) this cannot be.

We have thus shown that $\mathcal{G}$ only contains reflections of the type S_β, for example the $S_{\beta_1}, \ldots, S_{\beta_k}$ with $0 \leq \beta_1 < \cdots < \beta_k < \pi$. These are k different reflections, and therefore $S_{\beta_1} \circ S_{\beta_j}$ ($j = 2, \ldots, k$) $k - 1$ are different rotations. This shows $k \leq n$. Conversely:

---

[2] This is remarkable: Two non-trivial reflections as symmetry force a rotational symmetry.

$$S_{\beta_1}, R_\alpha \circ S_{\beta_1}, R_{2\alpha} \circ S_{\beta_1}, \ldots, R_{(n-1)\alpha} \circ S_{\beta_1}$$

are n different reflections, So it is $n \leq k$. It follows that $k = n$, and the reflections in $\mathcal{G}$ must be the movements $R_{j\alpha} \circ S_{\beta_1} = S_{\beta_1 + j\alpha/2}$ $(j = 0, \ldots, n-1)$.

It still needs to be proven that $\mathcal{G}$ and the symmetry group $\mathcal{G}'$ of Δ_n are equivalent. Choose any reflection S' in $\mathcal{G}'$. The elements of $\mathcal{G}$ are the $2n$ movements $\mathrm{Id}, R_\alpha, \ldots, R_{(n-1)\alpha}, S := S_{\beta_1}, R_\alpha \circ S, \ldots, R_{(n-1)\alpha} \circ S$. The mapping Φ from $\mathcal{G}$ to $\mathcal{G}'$ is defined as follows:

$$\mathrm{Id} \mapsto \mathrm{Id}, \; R_\alpha \mapsto R_\alpha, \; R_{j\alpha} \circ S \mapsto R_{j\alpha} \circ S'.$$

It is easy to see that Φ has all the required properties (bijective, group morphism, rotations on rotations, reflections on reflections). Consequently, the groups are equivalent.   □

For illustration, the symmetry groups for the letters of the German and Russian alphabets are listed below:

A B C D E F G H I J K L M N O P Q R S T U V W X Y Z

German. *gray*: trivial; *red*: D_1; *green*: C_2; *blue*: D_2

А Б В Г Д Е Ё Ж И Й К Л М Н О П Р

С Т У Ф Х Ц Ч Ш Щ Ъ Ы Ь Э Ю Я

Russian. *gray*: trivial; *red*: D_1; *green*: C_2; *blue*: D_2

And here are the symmetry groups of some traffic signs:

Trivial symmetry group. (Without text, the stop sign would have the D_8 as its symmetry group.)

Symmetry group D_1

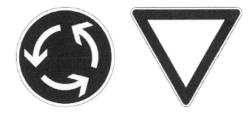

Symmetry groups C_3 and D_3

This traffic sign has an infinite symmetry group (many rotations and reflections)

Examples of the symmetry groups discussed here can also be found in nature: flowers, starfish, microorganisms, ...

   The finite symmetry groups are sometimes also called *rosette groups*: rosettes are round windows designed in an artistic way in architecture (for example at churches or town halls), which often show interesting symmetries.

Rosettes at church windows

A rosette from the Museum für angewandte Kunst in Vienna

For all these groups, *fundamental domains F* are easy to find. In the case of translations and glide reflections, one can choose a strip that is as wide as the translation distance and that is perpendicular to the translation direction. For the C_n or D_n, an angle space with the tip in the rotation center and the opening angle $2\pi/n$ (or π/n) is suitable.

F can be designed arbitrarily and then all movements of the group can be applied to it. However, you have to be careful at the edges of F. If it is reflected there, your picture should take this into account. For example, a semi-circle becomes a full circle. If the edge is rotated to another edge piece, however, the pieces should fit together. An arrowhead on one side could be continued by an end of an arrow at the other end.

3.3   The Subgroup of Translations

Let $\mathcal{G}$ be a discontinuous group of movements (cf. Definition 2.4.1) and $\mathbb{T}$ the subgroup of translations in $\mathcal{G}$. It plays an important role.

If $\mathbb{T}$ is trivial (i.e. only consists of Id), then we have already completely characterized the associated groups in proposition 3.2.3. There are two more possibilities for $\mathbb{T}$, which we describe in the next two propositions.

Proposition 3.3.1
Let $\mathcal{G}$ be a discontinuous group of movements. If $\mathbb{T}$ is nontrivial and the $\mathbf{a}$ for the $T_{\mathbf{a}} \in \mathcal{G}$ are each linearly dependent, then $\mathbb{T}$ is cyclic: There is a $T_{\mathbf{a}_0} \in \mathbb{T}$ so that $\mathbb{T} = \{T_{n\mathbf{a}_0} \mid n \in \mathbb{Z}\}$.

Proof Because of the assumption one can choose $\mathbf{a} \neq 0$ so that each translation $T \in \mathbb{T}$ has the form $T_{t\mathbf{a}}$ with a $t \in \mathbb{R}, t \neq 0$. The set W of occurring t is then a closed subgroup of $\mathbb{R}$ (otherwise $\mathbb{T}$ would not be discontinuous), and it follows that W can be written as $t_0 \mathbb{Z}$ for a suitable $t_0 \neq 0$. (Choose t_0 as the smallest strictly positive element of W.) Set $\mathbf{a}_0 := t_0 \mathbf{a}$.   □

In definition 2.3.3 we had already introduced Δ as the set of $\mathbf{a}$ with $T_{\mathbf{a}} \in \mathcal{G}$. We now assume that $\mathbb{T}$ is not trivial and that the above situation does not occur: There are therefore linearly independent vectors in Δ. We then define:

Definition 3.3.2
Two linearly independent vectors $\mathbf{a}_0$, $\mathbf{b}_0$ are called *a basis* of Δ if $\Delta = \{T_{n\mathbf{a}_0 + m\mathbf{b}_0} \mid n, m \in \mathbb{Z}\}$.

According to definition 2.3.3 we had already sketched two examples of Δ. In both cases, it is easy to specify different bases.

Examples of bases in translation grids

But how can you find bases? The following lemma provides a useful criterion:

Lemma 3.3.3

Let $\mathbf{a}_0$, $\mathbf{b}_0 \in \Delta$ be linearly independent vectors such that the generated parallelogram $P := \{s\mathbf{a}_0 + t\mathbf{b}_0 \mid 0 \leq s, t \leq 1\}$ contains no elements of Δ other than the vectors $\mathbf{0}, \mathbf{a}_0, \mathbf{b}_0, \mathbf{a}_0 + \mathbf{b}_0$ (these are the corners of P). Then $\mathbf{a}_0, \mathbf{b}_0$ is a basis of Δ.

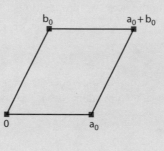

The parallelogram P

Proof Let $\Delta' := \{T_{n\mathbf{a}_0 + m\mathbf{b}_0} \mid n, m \in \mathbb{Z}\}$. We have $\Delta' \subset \Delta$, and we claim equality. For this, let $\mathbf{a} \in \Delta$. Since $\mathbf{a}_0$, $\mathbf{b}_0$ are linearly independent, we can write $\mathbf{a}$ as $s'\mathbf{a}_0 + t'\mathbf{b}_0$, where $s', t' \in \mathbb{R}$. If we choose $m, n \in \mathbb{Z}$ with $s' - m, t' - n \in [0, 1]$, then $\mathbf{a} - m\mathbf{a}_0 - n\mathbf{b}_0 \in P \cap \Delta$, so in $\{\mathbf{0}, \mathbf{a}_0, \mathbf{b}_0, \mathbf{a}_0 + \mathbf{b}_0\}$: It follows that $\mathbf{a} \in \Delta'$.   □

Lemma 3.3.4

$\mathcal{G}$ is again discontinuous, and Δ contains linearly independent vectors. Choose $\mathbf{a}_0 \in \Delta \setminus \{\mathbf{0}\}$ with minimal norm and then $\mathbf{b}_0$ with minimal norm in $\Delta \setminus \mathbb{Z}\mathbf{a}_0$. (Such vectors exist, because otherwise the group would not be discontinuous.) Then $\mathbf{a}_0, \mathbf{b}_0$ is a basis of Δ.

Proof The vectors are obviously linearly independent. As just mentioned, we denote with P the spanned parallelogram, and we claim that the conditions of the previous lemma are satisfied. To do this, we will show: If $\mathbf{x}$ is an arbitrary point in P, then

$$\min\{\|\mathbf{x}\|, \|\mathbf{x} - \mathbf{a}_0\|, \|\mathbf{x} - \mathbf{b}_0\|, \|\mathbf{x} - (\mathbf{a}_0 + \mathbf{b}_0)\|\} < \|\mathbf{b}_0\|. \tag{$*$}$$

We assume that $(*)$ has already been proven and choose an $\mathbf{x} \in \Delta \cap P$. If $\mathbf{x}$ lies on the line from $\mathbf{0}$ to $\mathbf{a}_0$, then $\mathbf{x} = \mathbf{0}$ or $\mathbf{x} = \mathbf{a}_0$ must hold, because otherwise $\mathbf{a}_0$ would not have been chosen with minimal norm. In all other cases, $\mathbf{x}$ and $\mathbf{a}_0$ are linearly independent, i. e. $\|\mathbf{x}\| \geq \|\mathbf{b}_0\|$ after choice of $\mathbf{b}_0$. Can $\|\mathbf{x} - \mathbf{a}_0\| < \|\mathbf{b}_0\|$ be? $\mathbf{x} - \mathbf{a}_0$ belongs to Δ, and therefore there is only no contradiction to the choice of $\mathbf{b}_0$ if $\mathbf{x} - \mathbf{a}_0 = \mathbf{0}$, i. e. $\mathbf{x} = \mathbf{a}_0$. Quite analogously, it follows from $\|\mathbf{x} - \mathbf{b}_0\| < \|\mathbf{b}_0\|$

or $\|\mathbf{x} - (\mathbf{a}_0 + \mathbf{b}_0)\| < \|\mathbf{b}_0\|$ that $\mathbf{x} = \mathbf{b}_0$ or $\mathbf{x} = \mathbf{a}_0 + \mathbf{b}_0$ holds. Together, it would therefore follow: $P \cap \Delta = \{\mathbf{0}, \mathbf{a}_0, \mathbf{b}_0, \mathbf{a}_0 + \mathbf{b}_0\}$, so the conditions of the previous lemma are satisfied.

For the proof of $(*)$, let $\mathbf{x} = s\mathbf{a}_0 + t\mathbf{b}_0 \in P$ be given with $0 \le s, t \le 1$. First, let $s + t \le 1$, $\mathbf{x}$ therefore lies in the triangle spanned by $\mathbf{0}, \mathbf{a}_0, \mathbf{b}_0$.

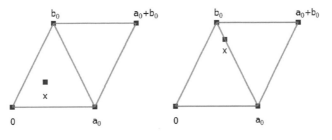

The cases $s + 1 < 1$ and $s + t = 1$

If even $s + t < 1$, then

$$\|\mathbf{x}\| = \|s\mathbf{a}_0 + t\mathbf{b}_0\| \le s\|\mathbf{a}_0\| + t\|\mathbf{b}_0\| \le (s + t)\|\mathbf{b}_0\| < \|\mathbf{b}_0\|.$$

In the case $s + t = 1$, $\mathbf{x}$ lies on the connecting line from $\mathbf{a}_0$ to $\mathbf{b}_0$. Since $\mathbf{a}_0, \mathbf{b}_0$ are linearly independent, it follows from the Cauchy-Schwarz inequality that $\|\mathbf{a}_0 - \mathbf{b}_0\| < \|\mathbf{a}_0\| + \|\mathbf{b}_0\|$. On the other hand, $\|\mathbf{a}_0 - \mathbf{b}_0\| = \|\mathbf{a}_0 - \mathbf{x}\| + \|\mathbf{x} - \mathbf{b}_0\|$, since distances add up on a line. If now $\|\mathbf{a}_0 - \mathbf{x}\|, \|\mathbf{x} - \mathbf{b}_0\| \ge \|\mathbf{b}_0\|$, then

$$2\|\mathbf{b}_0\| \le \|\mathbf{a}_0 - \mathbf{x}\| + \|\mathbf{b}_0 - \mathbf{x}\| = \|\mathbf{a}_0 - \mathbf{b}_0\| < \|\mathbf{a}_0\| + \|\mathbf{b}_0\|$$

and thus the contradiction $\|\mathbf{b}_0\| < \|\mathbf{a}_0\|$.

In the case $1 \le s + t \le 2$, $\mathbf{x}$ lies in the triangle spanned by $\mathbf{a}_0, \mathbf{b}_0, \mathbf{a}_0 + \mathbf{b}_0$, and one can argue analogously. □

Proposition 3.3.5
Suppose that $\mathcal{G}$ is discontinuous, and that there are translations in linearly independent directions. Then there is a basis $\mathbf{a}_0, \mathbf{b}_0$ of Δ.

Proof This follows from the two preceding lemmas. □

Corollary 3.3.6
Suppose that $\mathcal{G}$ is a discontinuous group, and that there is a rotation R_α in $\mathcal{G}$ with $\alpha \ne \pi$. Then there is a basis $\mathbf{a}_0, \mathbf{b}_0$, where $\mathbf{a}_0$ has minimal norm and $\|\mathbf{a}_0\| = \|\mathbf{b}_0\|$ holds.

Proof One constructs $\mathbf{a}_0$ as above. One can then choose $\mathbf{b}_0 = R_\alpha \mathbf{a}_0$ due to proposition 2.3.4. These are due to the condition on α linearly independent vectors with the same norm.

$\square$

3.4   The 7 Frieze Groups

In this section we study so-called *Frieze groups*, where $\mathbb{T}$ is generated by a single translation. A *frieze* is an architectural decorative element, with which house facades are beautified. But one must continue the image in one's thoughts in both directions to infinity.

$$\mathsf{F7\ F7\ F7\ F7\ F7\ F7}$$
$$\mathsf{LJ\ LJ\ LJ\ LJ\ LJ\ LJ}$$

A frieze

Similar symmetries can be discovered in fences, stairs, etc., also there one must interpret the continuation to infinity appropriately.

More examples of friezes

For the rest of the chapter, let $\mathcal{G}$ be a discontinuous group of motions, for which $\mathbb{T}$ is the cyclic group generated by a translation $T_{\mathbf{a}_0}$. We first show that $\mathcal{G}$ can only contain "very simple" motions:

Lemma 3.4.1
(i)   If $\mathcal{G}$ contains a rotation $R_{\alpha, \mathbf{x}_0}$, then $\alpha = \pi$.
(ii) If $\mathcal{G}$ contains a reflection, then the reflection axis is in the direction of $\mathbf{a}_0$ or in the orthogonal direction. In the first or second case, the reflection is called *of type 1* or *of type 2*.
(iii) If $\mathcal{G}$ contains a glide reflection, then the reflection axis is in the direction of $\mathbf{a}_0$.

Proof (i) Suppose that $\mathcal{G}$ contains a rotation $\mathbf{x} \mapsto \mathbf{a} + R_\alpha \mathbf{x}$. We conjugate $T_{\mathbf{a}_0}$ with this map and, due to proposition 2.3.2, obtain that $\mathcal{G}$ must also contain the translation $T_{R_\alpha \mathbf{a}_0}$. $T_{R_\alpha \mathbf{a}_0}$ is only then contained in $\{T_{n\mathbf{a}_0} \mid n \in \mathbb{Z}\}$ if $n = \pi$.

(ii) Suppose that $S : \mathbf{x} \mapsto \mathbf{s} + S_\beta \mathbf{x}$ is a reflection (so $S_\beta \mathbf{s} = -\mathbf{s}$ applies). It follows again (after conjugating $T_{\mathbf{a}_0}$ with S) from proposition 2.3.2 that $\mathcal{G}$ must contain the translation $T_{S_\beta \mathbf{a}_0}$. But that is only possible if $S_\beta \mathbf{a}_0 = \mathbf{a}_0$ or $S_\beta \mathbf{a}_0 = -\mathbf{a}_0$, that is, if the direction of the mirror axis is parallel to $\mathbf{a}_0$ or perpendicular to it.

(iii) Let $T : \mathbf{x} \mapsto \mathbf{b} + \mathbf{s} + S_\beta \mathbf{x}$ be a glide reflection: $S_\beta \mathbf{s} = -\mathbf{s}$, $S_\beta \mathbf{b} = \mathbf{b}$, $\mathbf{b} \neq \mathbf{0}$. T^2 is the translation $T_{2\mathbf{b}}$, $2\mathbf{b}$ is therefore of the form $n\mathbf{a}_0$, where $n \neq 0$. From $S_\beta \mathbf{b} = \mathbf{b}$ and $n \neq 0$ it then follows that $S_\beta \mathbf{a}_0 = \mathbf{a}_0$. □

If there is a reflection S of type 1 in $\mathcal{G}$, of course there are also glide reflections (all $T_{n\mathbf{a}_0} \circ S$ with $n \in \mathbb{Z}, n \neq 0$). But it can also happen that there are glide reflections that *do not* have this form. We will call them *proper glide reflections*, they have already been introduced in proposition 2.3.6 [3].

For $\mathcal{G}$, the following four questions can be answered with $j = $ "yes" or $n = $ "n" (In the German version we used "j" for "ja" and "n" for "nein".):

Is there a reflection of type 1 in $\mathcal{G}$?
Is there a reflection of type 2 in $\mathcal{G}$?
Is there a rotation in $\mathcal{G}$?
Is there a proper glide reflection in $\mathcal{G}$?

There are $2^4 = 16$ possible answers, but only 7 can occur, and they will allow a complete classification.

Proposition 3.4.2

(i) The 16 possible answers to the 4 questions are listed below. The ones marked with a "*" cannot occur:

nnnn	*nnnj*	*nnjn*	*nnjj**	*njnn*	*njnj**	*njjn**	*njjj*
jnnn	*jnnj**	*jnjn**	*jnjj**	*jjnn**	*jjnj**	*jjjn*	*jjjj**

(ii) For each Frieze group $\mathcal{G}$, exactly one of the 7 possible cases occurs.

Proof (i) *nnjj**: If a rotation and a glide reflection with mirror axis G (according to Lemma 3.4.1 in the direction of $\mathbf{a}_0$) are contained, then also a reflection or glide reflection in an axis orthogonal to G, because a R_π-rotation rotates the mirror axis by $\pi/2$ (cf. Lemma 2.2.4 (vi)). But this cannot be a glide reflection with a shift different from zero, because this would result in a translation in a direction

---

[3] Please note: It should be emphasized again that it depends on $\mathcal{G}$ whether a glide reflection is proper or not.

orthogonal to $\mathbf{a}_0$ after being applied twice. But there is no such thing. So the product is a reflection of type 2.

njnj*: The product of two reflections or glide reflections with orthogonal mirror axes is a R_π-rotation. *njnj* would therefore imply *njjj*.

njjn*: The product of a reflection with mirror line G and a rotation is a reflection in a line orthogonal to G. *njjn* would therefore imply *jjjn*.

jnnj*: In the previous proof part, it was explained that the glide reflection cannot be proper.

jnjn*, *jnjj**: See the argument for *njjn**.

jjnn*, *jjnj**: The product of reflections in orthogonal lines is a R_π-rotation.

jjjj*: See the argument for *jnnj**.

(ii) This is clear, since it is a complete case distinction □

We will now show that for each of the 7 possible cases, up to equivalence, exactly one G is possible. In this sense, one can then say that there are exactly 7 Frieze groups. We treat the 7 cases in the following order:

- *nnnn*: There are only translations.
- *jnnn*: There is a type-1 reflection, but no type-2 reflections, rotations, and proper glide reflections.
- *njnn*: There is a type-2 reflection, but no type-1 reflections, rotations, and proper glide reflections.
- *nnnj*: There is a proper glide reflection, but no type-1 and type-2 reflections and no rotations.
- *nnjn*: There is a rotation, but neither reflections nor glide reflections.
- *jjjn*: There is a rotation, type-1 and type-2 reflections, but no proper glide reflection.
- *njjj*: There is a rotation, a proper glide reflection, and a type-2 reflection, but no type-1 reflection.

It is then clear that groups of motion that belong to different lines cannot be equivalent. We therefore only have to show for the proof of the statement "There are exactly seven Frieze groups, except for equivalence":

- An example can be given for each of the seven cases. This will be quite simple: We will give an image for which the symmetry group has the corresponding properties.
- If two movement groups fulfill the same conditions (for example *nnjn*), they are equivalent.

To prove the second statement, we show that in each case a complete catalog of the movements contained and the group table of the links can be set up for the presented group. If two corresponding groups are given, the associated catalogs lead to a bijection. Translations are mapped to translations, reflections to reflections, etc., and since the group tables are identical, it is also a group isomorphism.

Sometimes it will be convenient to conjugate the group before investigating it. If, for example, one knows that a rotation is included, one conjugates so that a rotation by the same angle around the origin belongs to $\mathcal{G}$: This is due to Lemma 3.1.2 (iii). This preparatory step is legitimate for our investigations, because conjugate subgroups are equivalent.

3.4.1 $\mathcal{F}_1$: Only Translations (*nnnn*)

This is the simplest case: There is an $\mathbf{a}_0 \neq \mathbf{0}$ so that

$$\mathcal{G} = \mathbb{T} = \{T_{n\mathbf{a}_0} \mid n \in \mathbb{Z}\}.$$

$\mathcal{G}$ therefore contains no rotations, reflections or glide reflections.

The official name $\mathcal{F}_1$. (For the Frieze groups and the plane groups treated in the next chapter, there are names that are used universally. They go back to the Hungarian mathematician Fejes Tóth.)

There are lines that are invariant under $\mathcal{G}$ Any line that is parallel to $\mathbf{a}_0$ is invariant.

Two groups of this type are equivalent Such a $\mathcal{G}$ consists of the $T_n := T_{n\mathbf{a}_0}$ with $n \in \mathbb{Z}$, and the group operations are given by $T_n \circ T_m := T_{n+m}$. If another candidate is the set of $T'_n = T_{n\mathbf{a}'_0}$ with the same group table, then $T_n \mapsto T'_n$ provides a group isomorphism, and that is even an equivalence, since translations are mapped to translations and other types of movements do not occur.

There is an example, and it is even a symmetry group We have already met the following example in Sect. 3.1.

FFFFFFFFFFF

The symmetry group is of type $\mathcal{F}_1$

Visualization In order to visualize possible translations, we will use small circles (the mutual distance corresponds to the vector $\mathbf{a}_0$). Note that the origin could be drawn at any point, since only the displacement matters. In the following sections it will always be "suitably" placed, for example in a rotation center or in a mirror axis.

O   O   O   O   O   O   O   O   O   O   O

Only translations

There is an example, and it is even a symmetry group

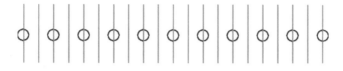

In the symmetry group there are only type-2 reflections and translations

Visualization We have chosen $\mathbf{a}_0$ in the direction of the x-axis. The mirror axes are then perpendicular and parallel at a distance $\|\mathbf{a}_0\|/2$.

Only reflections of type 2

3.4.4 $\mathcal{F}_1^3$: Proper Glide Reflections (*nnnj*)

The official name $\mathcal{F}_1^3$.

The analysis We already know (Lemma 3.4.1), that all glide mirror axes point in direction $\mathbf{a}_0$. Now let $G_1 : \mathbf{x} \mapsto \mathbf{s}_1 + \mathbf{b}_1 + S_\beta \mathbf{x}$, $G_2 : \mathbf{x} \mapsto \mathbf{s}_2 + \mathbf{b}_2 + S_\beta \mathbf{x}$ be glide reflections in $\mathcal{G}$. It is $S_\beta \mathbf{b}_1 = \mathbf{b}_1$, $S_\beta \mathbf{s}_1 = -\mathbf{s}_1$ and $S_\beta \mathbf{b}_2 = \mathbf{b}_2$, $S_\beta \mathbf{s}_2 = -\mathbf{s}_2$. Also $S_\beta \mathbf{a}_0 = \mathbf{a}_0$. We notice that G_2^{-1} is the mapping $\mathbf{x} \mapsto \mathbf{s}_2 - \mathbf{b}_2 + S_\beta \mathbf{x}$.

$G_1 \circ G_2^{-1}$ then has the form $\mathbf{x} \mapsto \mathbf{b}_1 + \mathbf{s}_1 - \mathbf{b}_2 - \mathbf{s}_2 + \mathbf{x}$. It is a translation, and thus $\mathbf{b}_1 + \mathbf{s}_1 - \mathbf{b}_2 - \mathbf{s}_2 = n\mathbf{a}_0$ for a $n \in \mathbb{Z}$.

Apply S_β to this equation and subtract the equation from it, then $-(\mathbf{s}_1 - \mathbf{s}_2) - (\mathbf{s}_1 - \mathbf{s}_2) = -2(\mathbf{s}_1 - \mathbf{s}_2) = \mathbf{0}$ follows, so it is $\mathbf{s}_1 = \mathbf{s}_2$, and thus $\mathbf{b}_1 - \mathbf{b}_2 = n\mathbf{a}_0$.

Now fix a proper glide reflection $G : \mathbf{x} \mapsto \mathbf{s} + \mathbf{b} + S_\beta \mathbf{x}$ in $\mathcal{G}$. It is $G^2 = T_{2\mathbf{b}}$, and this translation can be written as $T_{n\mathbf{a}_0}$. It is necessary that $n = 2k + 1$ is odd, because in the case $n = 2k$ would be $\mathbf{b} = k\mathbf{a}_0$, and $T_{-k\mathbf{a}_0} \circ G$ would be a type-1 reflection in $\mathcal{G}$.

So it is $n = 2k + 1$. Set

$$\hat{G} := G \circ T_{-k\mathbf{a}_0} \circ G : \mathbf{x} \mapsto \mathbf{s} + \hat{\mathbf{b}} + S_\beta \mathbf{x},$$

with $\hat{\mathbf{b}} = \mathbf{a}_0/2$. We have shown: All other glide reflections have the form $\mathbf{x} \mapsto n\mathbf{a}_0 + \mathbf{s} + \hat{\mathbf{b}} + S_\beta \mathbf{x}$. Conversely, this also applies: For arbitrary $n \in \mathbb{Z}$, $T_{n\mathbf{a}_0} \circ \hat{G} : \mathbf{x} \mapsto n\mathbf{a}_0 + \mathbf{s} + \hat{\mathbf{b}} + S_\beta \mathbf{x}$ is a real glide reflection. (It is a glide reflection, and all glide reflections in $\mathcal{G}$ should be proper according to the assumption.)

If you set $T_n := T_{n\mathbf{a}_0}$ and $G_n := T_{n\mathbf{a}_0} \circ \hat{G}$, $\mathcal{G}$ therefore consists of the following movements:

translations	T_n with $n \in \mathbb{Z}$
mirror reflections, type 1	$\emptyset$
mirror reflections, type 2	$\emptyset$
π-rotations	$\emptyset$
glide reflections	G_n with $n \in \mathbb{Z}$ (proper)

The group operations:

$$T_n \circ T_m := T_{n+m}, \; T_n \circ G_m = G_m \circ T_n = G_{n+m}, \; G_n \circ G_m = T_{n+m+1}.$$

There are lines that are left invariant under $\mathcal{G}$ The glide reflection axis of $\hat{G}$ is the uniquely determined line that is left invariant under all $T \in \mathcal{G}$.

Two groups of this type are equivalent $\mathcal{G}$ and $\mathcal{G}'$ are groups of the type considered here. Choose $T_{\mathbf{a}_0} \in \mathcal{G}$ or $T_{\mathbf{a}_0'} \in \mathcal{G}'$ so that the translations in $\mathcal{G}$ or $\mathcal{G}'$ are just the $T_{n\mathbf{a}_0}$ or the $T_{n\mathbf{a}_0'}$. Then look for glide reflections $\hat{G} \in \mathcal{G}$ or $\hat{G}' \in \mathcal{G}'$ with $\hat{G}^2 = T_{\mathbf{a}_0}$ or $(\hat{G}')^2 = T_{\mathbf{a}_0'}$ as described above.

   Map $\mathcal{G}$ bijectively to $\mathcal{G}'$ by $T_{n\mathbf{a}_0} \mapsto T_{n\mathbf{a}_0'}$, $T_{n\mathbf{a}_0} \circ \hat{G} \mapsto T_{n\mathbf{a}_0'} \circ \hat{G}'$. Because the group calculation rules are the same in both groups, a group isomorphism between $\mathcal{G}$ and $\mathcal{G}'$ is induced. And that's an equivalence, because translations are mapped to translations and glide reflections are mapped to glide reflections.

There is an example, and it is even a symmetry group

In the symmetry group there are only proper glide reflections and translations

Visualization For glide reflections, the mirror axis is represented by a dashed line, and the feed is marked by short transverse strokes.

Only glide reflections

3.4.5 $\mathcal{F}_2$: Only Rotations (*nnjn*)

The official name $\mathcal{F}_2$.

The analysis $\mathcal{G}$ contains a rotation by the angle π. If we conjugate $\mathcal{G}$ appropriately, we can assume that $R_\pi \in \mathcal{G}$, because two rotations by the same angle are conjugate. This does not change our classification problem, because conjugate groups are equivalent (see Sect. 3.1).

R_π has the rotation center $\mathbf{0}$. All $T_{n\mathbf{a}_0} \circ R_\pi$ ($n \in \mathbb{Z}$) also lie in $\mathcal{G}$, and these rotations have the rotation centers $n\mathbf{a}_0/2$ (proposition 2.3.2). There are no more rotations: If $R : \mathbf{x} \mapsto \mathbf{a} + R_\pi \mathbf{x}$ is a rotation in $\mathcal{G}$, then $R \circ R_\pi$ is the translation $T_\mathbf{a}$, and therefore $\mathbf{a}$ has the form $n\mathbf{a}_0$.

Together this means that—with $T_n := T_{n\mathbf{a}}$ and $R_n : \mathbf{x} \mapsto n\mathbf{a}_0 + R_\pi \mathbf{x}$—the group $\mathcal{G}$ consists of the following elements[5]:

translations	T_n with $n \in \mathbb{Z}$
mirror reflections, type 1	$\emptyset$
mirror reflections, type 2	$\emptyset$
π-rotations	R_n with $n \in \mathbb{Z}$
glide reflections	$\emptyset$

The group operations are:

$$T_n \circ T_m = T_{n+m}, \; T_n \circ R_m = R_{n+m}, \; R_n \circ T_m = R_{n-m}, \; R_n \circ R_m = T_{n-m}.$$

There are lines that are left invariant under $\mathcal{G}$ The only invariant line is the line through the rotation centers. If we take $R_\pi \in \mathcal{G}$, then that is the line through $\mathbf{0}$ in the direction of $\mathbf{a}_0$.

Two groups of this type are equivalent As in the previous examples, we argue here.

There is an example, and it is even a symmetry group

F⅃ F⅃ F⅃ F⅃ F⅃ F⅃

The symmetry group is the $\mathcal{F}_2$

Visualization We denote rotations by a diamond.

◎ ◇ ◎ ◇ ◎ ◇ ◎ ◇ ◎ ◇ ◎ ◇ ◎ ◇ ◎ ◇ ◎ ◇ ◎

Rotation centers and translations for the $\mathcal{F}_2$

3.4.6 $\mathcal{F}_2^1$: Rotations, Type-1 and Type-2 Reflections (*jjjn*)

The official name $\mathcal{F}_2^1$.

The analysis As in the previous subsection, we begin by conjugating to achieve that R_π belongs to $\mathcal{G}$: This will simplify the investigations. Further, let $S : \mathbf{x} \mapsto \mathbf{s} + S_\beta \mathbf{x}$ be an arbitrary type-1 reflection. Then $S_\beta \mathbf{s} = -\mathbf{s}$, and $S_\beta \mathbf{a}_0 = \mathbf{a}_0$. We claim that $\mathbf{s} = \mathbf{0}$ holds.

---

[5] It may be necessary to conjugate beforehand to achieve $R_\pi \in \mathcal{G}$.

To this end, we first consider the reflection $S' := R_\pi \circ S \circ R_\pi : \mathbf{x} \mapsto -\mathbf{s} + S_\beta \mathbf{x}$ and then the product $S \circ S' : \mathbf{x} \mapsto \mathbf{s} + S_\beta(-\mathbf{s} + S_\beta \mathbf{x}) = 2\mathbf{s} + \mathbf{x}$. This is a translation, so that $2\mathbf{s} = n\mathbf{a}_0$ for a suitable $n \in \mathbb{Z}$. It follows that $-2\mathbf{s} = 2S_\beta \mathbf{s} = nS_\beta \mathbf{a}_0 = n\mathbf{a}_0 = 2\mathbf{s}$, i.e., $\mathbf{s} = \mathbf{0}$ as claimed.

Therefore $S := S_\beta \in \mathcal{G}$, and from Sect. 3.4.2 we know that there is only one type-1 reflection.

We now consider $S^* := S \circ R_\pi$. This is a type-2 reflection, because the axis of reflection is perpendicular to the axis of reflection of S. Together with the results from Sects. 3.4.3 and 3.4.5 this shows that $\mathcal{G}$ consists exactly of the following movements:

translations	T_n with $n \in \mathbb{Z}$
mirror reflections, type 1	S
mirror reflections, type 2	$S_n^* := T_n \circ S^*$ with $n \in \mathbb{Z}$
π-rotations	$R_n := T_n \circ R_\pi$ with $n \in \mathbb{Z}$
glide reflections	$G_n := T_n \circ S$ with $n \in \mathbb{Z}, n \neq 0$ (improper)

The group calculation rules are listed in the previous sections or can be easily derived from the formulas proved in Sect. 2.2; for example $R_n \circ S_m = S_{n-m}^*$ etc.

There are lines that are invariant under $\mathcal{G}$ There is exactly one line that is invariant: The mirror axis of the type-1 reflection.

Two groups of this type are equivalent The same argument as in the previous cases leads to the goal.

There is an example, and it is even a symmetry group

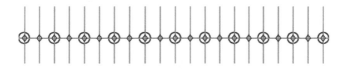

Symmetry group is the $\mathcal{F}_2^1$

Visualization

Rotations and reflections of type 1 and 2

3.4.7 $\mathcal{F}_2^2$: Proper Glide Reflections, Type-2 Reflections, and Rotations (njjj)

The official name $\mathcal{F}_2^2$.

The analysis Again, R_π should belong to $\mathcal{G}$. We choose an arbitrary proper glide reflection $G : \mathbf{x} \mapsto \mathbf{s} + \mathbf{b} + S_\beta \mathbf{x}$ in $\mathcal{G}$. (here $S_\beta \mathbf{b} = \mathbf{b}, S_\beta \mathbf{s} = -\mathbf{s}, S_\beta \mathbf{a}_0 = \mathbf{a}_0$.) $R_\pi \circ G \circ R_\pi$ is the glide reflection $G' : \mathbf{x} \mapsto \mathbf{s} - \mathbf{b} + S_\beta \mathbf{x}$, and with $G \circ G'$ we get the translation $T_{2\mathbf{s}}$. Since it lies in $\mathcal{G}$ it follows that $\mathbf{s} = \mathbf{0}$. $\mathbf{s} = \mathbf{0}$.

Thus $G : \mathbf{x} \mapsto \mathbf{b} + S_\beta \mathbf{x}$ in $\mathcal{G}$, and as in Sect. 3.4.4 we may—after multiplication with a suitable translation—assume that $2\mathbf{b} = \mathbf{a}_0$ holds.

Set $S := G \circ R_\pi$, which is the type-2 reflection $\mathbf{x} \mapsto \mathbf{b} + S_{\beta+\pi/2}$. From the considerations in Sects. 3.4.3, 3.4.4 and 3.4.5 it now follows how all elements from $\mathcal{G}$ look:

translations	T_n with $n \in \mathbb{Z}$
mirror reflections, type 1	$\emptyset$
mirror reflections, type 2	$T_n \circ S$ with $n \in \mathbb{Z}$
π-rotations	$\emptyset$
glide reflections	$T_n \circ G$ with $n \in \mathbb{Z}$ (proper)

The group operations are again easily derived from the formulas in Sect. 2.2.

There are lines that are left invariant under $\mathcal{G}$ The line of reflection of the glide reflections is the only invariant line.

Two groups of this type are equivalent This can be shown completely analogously to the argumentation in the previous sections.

There is an example, and it is even a symmetry group

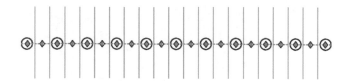

Symmetry group is the $\mathcal{F}_2^2$

Visualization

Proper glide reflections, Type-2 reflections, and rotations

3.4.8   Summary

The above results can be reformulated as follows:

- Let $\mathcal{G}_0$ be the group of all movements on $\mathbb{R}^2$. Consider the subset of all discontinuous subgroups in the set of all subgroups. This set decomposes into 3 disjoint parts, depending on whether the respective subgroup of translations is trivial or generated by one or by two independent translations.
- Consider in particular those discontinuous subgroups in which the translations are generated by a single translation. Introduce an equivalence relation: Equivalence means the existence of a group isomorphism that converts translations (rotations, reflections, glide reflections) into translations (rotations, reflections, glide reflections).
- Then there are exactly seven equivalence classes, and in each lies a symmetry group.
- In each case there is a line that is carried into itself under all the movements of the group[6].

3.4.9   Classification: A Test

The author of this book has generated seven friezes below using an image of himself. Dear readers, you are cordially invited to classify these friezes: to which of the 7 types does each one belong? The solutions can be found at the end of Sect. 3.4.10.

Figure 1

---

[6]Alternatively, one could have defined: A Frieze group is a discontinuous group of movements for which there is a line that is invariant under all the movements of the group.

Figure 2

Figure 3

Figure 4

Figure 5

Figure 6

Figure 7

3.4.10 Hints for Artists

How can one create a pattern whose symmetry group has one of the seven possible types of frieze groups?

One first fixes some frieze group $\mathcal{G}$. Then one proceeds in two steps. First, one must choose a fundamental domain F to $\mathcal{G}$, that is, a closed subset of the $\mathbb{R}^2$, so that first $\bigcup_{T\in\mathcal{G}} T(F) = \mathbb{R}^2$ and that secondly $T(F)$ and F for $T \in \mathcal{G}, T \neq$ Id intersect only in boundary points.

We want to assume that F is limited by line segments $G_1, \ldots, G_l$. Then one must analyze which group operations can transform a G_i into a G_j: a translation? Rotation? Reflection? Glide reflection?

Then F can be decorated arbitrarily. There are no restrictions inside of F, but one must be careful at the edge. If, for example, a $P \in G_i$ is transformed into a $Q \in G_j$ through a *translation* or a *rotation*, then the image in P must be continued in Q. If, for example, the end of an arrow is seen in P, then the arrowhead must appear in Q.

But if $P \in G_i$ is transformed into $Q \in G_j$ through a *reflection*, then at P one half of a symmetrical pattern (for example, half of a smiling face) must be seen, whose other half is to be found at Q.

The solutions to the classification problem in Sect. 3.4.9

Figure 1:$\mathcal{F}_2$ Figure 2:$\mathcal{F}_2^2$ Figure 3:$\mathcal{F}_1^2$ Figure 4:$\mathcal{F}_1^1$
Figure 5:$\mathcal{F}_1^3$ Figure 6:$\mathcal{F}_1$ Figure 7:$\mathcal{F}_2^1$

3.5 The 17 Plane Crystal Groups

Similarly to the Frieze groups, we now want to derive a complete characterization of those discrete movement groups of the plane, in which there are *two* linear independent translation directions. The proofs are significantly more involved than in the case of finite groups and in the case of Frieze groups. A fundamental role is played by the so-called *crystallographic restriction*, which we prove in the first subsection: In the case under consideration, only "few" rotation angles for rotations in the group are possible.

Then various cases are systematically investigated: What can be said if there are only translations and reflections? What changes if 2-rotations are also possible? Or 3-, 4- or 6-rotations? The key to characterization is a careful analysis of the properties of the translation grids in each of these cases.

3.5.1   The Crystallographic Restriction

We now prove a central result, through which a concise classification of the discontinuous groups is possible at all. It states that in such groups rotations $\mathbf{a} + R_\alpha$ are only possible for "very few α". As a preparation, we show:

Lemma 3.5.1

Let $\mathcal{G}$ be a group of movements, $\alpha \in \mathbb{R} \setminus 2\pi\mathbb{Z}$ and $\mathbf{x}', \mathbf{y}'$ rotation centers to α-rotations in $\mathcal{G}$. The maps $S : \mathbf{x} \mapsto \mathbf{a} + R_\alpha\mathbf{x}$ and $T : \mathbf{x} \mapsto \mathbf{a}' + R_\alpha\mathbf{x}$ therefore belong to $\mathcal{G}$, where $\mathbf{a} := \mathbf{x}' - R_\alpha\mathbf{x}'$ and $\mathbf{a}' := \mathbf{y}' - R_\alpha\mathbf{y}'$.
If T rotates with center $\mathbf{y}'$ by α, then T^{-1} must rotate with center $\mathbf{y}'$ by $-\alpha$, i. e. $T^{-1}\mathbf{x} = (\mathbf{y}' - R_{-\alpha}\mathbf{y}') + R_{-\alpha}\mathbf{x}$.

(i)   Also $S\mathbf{y}'$ is a α-rotation center of a rotation in $\mathcal{G}$.
(ii) $T^{-1}\mathbf{x}'$ is a α-rotation center for a rotation in $\mathcal{G}$.

Proof (i) $R := S \circ T \circ S^{-1}$ belongs to $\mathcal{G}$. We claim that R is an α-rotation with rotation center $S\mathbf{y}'$.

First we calculate R explicitly. If one uses the definitions for S and T, then $R : \mathbf{x} \mapsto \mathbf{a} - R_\alpha\mathbf{a} + R_\alpha\mathbf{a}' + R_\alpha\mathbf{x}$ follows, so it is an α-rotation. We have to show that

$$S\mathbf{y}' - R_\alpha S\mathbf{y}' = \mathbf{a} - R_\alpha\mathbf{a} + R_\alpha\mathbf{a}'.$$

Really we have

$$
\begin{aligned}
S\mathbf{y}' - R_\alpha S\mathbf{y}' &= \mathbf{x}' - R_\alpha\mathbf{x}' + R_\alpha\mathbf{y}' - R_\alpha(\mathbf{x}' - R_\alpha\mathbf{x}' + R_\alpha\mathbf{y}') \\
&= \mathbf{x}' - 2R_\alpha\mathbf{x}' + R_\alpha\mathbf{y}' + R_{2\alpha}\mathbf{x}' - R_{2\alpha}\mathbf{y}' \\
&= \mathbf{x}' - R_\alpha\mathbf{x}' - R_\alpha(\mathbf{x}' - R_\alpha\mathbf{x}') + R_\alpha(\mathbf{y}' - R_\alpha\mathbf{y}') \\
&= \mathbf{a} - R_\alpha\mathbf{a} + R_\alpha\mathbf{a}'.
\end{aligned}
$$

(ii) This follows from (i), since α-rotation centers are also $-\alpha$-rotation centers.   □

Proposition 3.5.2

Let $\mathcal{G}$ be a group of movements, and α is of the form $2\pi/n$ for a $n \in \{2, 3, \ldots\}$. If $n = 5$ or $n > 6$, then: If $\mathbf{x}', \mathbf{y}'$ α-rotation centers for rotations in $\mathcal{G}$ with $\mathbf{x}' \neq \mathbf{y}'$, then there are α-rotation centers $\mathbf{x}'', \mathbf{y}''$ with $\mathbf{x}'' \neq \mathbf{y}''$ for rotations in $\mathcal{G}$ with $\|\mathbf{x}'' - \mathbf{y}''\| < \|\mathbf{x}' - \mathbf{y}'\|$: There is therefore no smallest distance for α-rotation centers.

Proof Choose rotation centers $\mathbf{x}'$, $\mathbf{y}'$, the distance is l. Draw over the segment from $\mathbf{x}'$ to $\mathbf{y}'$ more segments of length l: at $\mathbf{x}'$ at an angle of α clockwise, at $\mathbf{y}'$ at an angle of α counterclockwise. The endpoints of the newly drawn routes are the points x'', y'' of the previous lemma:

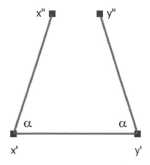

In *red*: x' and y'; in *blue*: x'' and y''

And its distance is smaller than l, if $n = 5$ or $n > 6$. This is clear from the following pictures: Shown are the cases $n = 5$ and as an example the cases $n = 7$ and $n = 12$. □

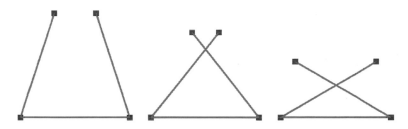

There is no smallest distance of the rotation centers, if $n = 5$ or $n > 6$

There remain the cases $n = 2, 3, 4, 6$. The transition from x', y' to x'', y'' *not* leads to a contradiction. This is clear for $n = 2$. In the case $n = 3$, the distance is not reduced, in the case $n = 4$ it remains the same, and in the case $n = 6$, $x'' = y''$ holds.

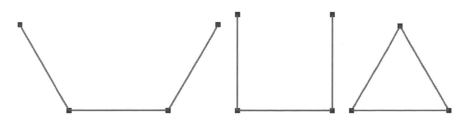

The transition from x', y' to x'', y'' in the cases $n = 3, n = 4$ and $n = 6$

Proposition 3.5.3 (Crystallographic Restriction)
Let G be a discontinuous group of movements. If there is an α-rotation in G, then $\alpha \in \{\pi, 2\pi/3, 2\pi/4, 2\pi/6\}$. So only 2-, 3-, 4 - and 6-rotations are possible.

Proof This follows directly from the previous proposition: If G is discontinuous, then with Δ also $\Delta_\alpha = (\mathrm{Id} - R_\alpha)^{-1}\Delta$ is discrete, and consequently the smallest distance between the rotation centers is realized. $\qquad\square$

Proposition 3.5.4
Let G be a discontinuous group of movements.

(i)   If G contains a 4-rotation, then G does not contain any 3- or 6-rotations.
(ii) If G contains a 3-rotation or a 6-rotation, then G does not contain any 4-rotations.

Proof Note only: If there are 4- and 3-rotations (or 4- and 6-rotations), then there are also 12-rotations. And something like this cannot be in discontinuous groups because of the previous proposition. $\qquad\square$

3.5.2   Translations, Reflections: 4 Groups

We now begin with a systematic discussion of the plane discontinuous groups of movements. In this first subsection we treat groups that only contain translations and possibly reflections: There are thus no rotations. The structure is the same in all cases. In addition to general further preparations, we treat the following points:

- What property does the group have?
- How is it officially designated?
- What do you need to know to be able to successfully analyze this case?
- The property uniquely determines the group up to equivalence.
- There are examples that are even symmetry groups.
- How can groups of this type be visualized?

The strategy corresponds to that used in the case of Frieze groups:

- After choosing an appropriate basis of the translation grid, compile a catalog of the movements contained in the group.
- Calculate the group table.

If two groups with the same properties are given, use the catalog to define a bijection. Since the group tables match, a group isomorphism has been created, and since translations are mapped to translations, reflections to reflections, etc., even an equivalence is present.

Sometimes it will also be convenient, as in the case of the Frieze groups, to conjugate first. For example, if you know that a 3-rotation is included, you may o. B. d. A. assume that $R_{2\pi/3} \in \mathcal{G}$.

In the end it will be shown that there are exactly 17 pairwise different possibilities for discrete movement groups with two linearly independent translation directions.

Group 1

The property There are only translations.

The official notation $\mathcal{W}_1$ or $p1$.

The analysis With the notations from Sect. 3.3 we have $\mathcal{G} = \mathbb{T}$. If one sets $T_{m,n} := T_{m\mathbf{a}_0 + n\mathbf{b}_0}$ for $m, n \in \mathbb{Z}$ (where $\mathbf{a}_0, \mathbf{b}_0$ denotes a basis of Δ), then $\mathcal{G}$ consists of the following movements:

translations	$T_{m,n}$ with $m, n \in \mathbb{Z}$
mirror reflections	none
rotations	none
glide reflections	none

And the group table looks, as is easily seen, as follows:

$$T_{m_1,n_1} \circ T_{m_2,n_2} = T_{m_1+m_2,n_1+n_2}.$$

Two groups of this type are equivalent If G, G' are groups with the property considered here, then choose bases $\mathbf{a}_0, \mathbf{b}_0$ and $\mathbf{a}_0', \mathbf{b}_0'$ in the corresponding translation grids. $T_{m\mathbf{a}_0 + n\mathbf{b}_0} \mapsto T_{m\mathbf{a}_0' + n\mathbf{b}_0'}$ is then a group isomorphism from $\mathcal{G}$ to $\mathcal{G}'$. The groups are then also equivalent, since translations are mapped to translations.

There is an example, and it is even a symmetry group

In the symmetry group there are only translations

Visualization As with Frieze groups, we use small red circles for the elements of Δ, lines for mirror axes, dashed lines for glide mirror axes, and a rhombus for rotation centers of R_π rotations. Later we will need triangles (for 3 rotations), squares (for 4 rotations), and hexagons (for 6 rotations). Actually, one should then nest a hexagon, a rhombus, and a triangle, for example, because centers for 6 rotations are also centers for 2 and 3 rotations. But for the sake of clarity, we will only draw the hexagon.

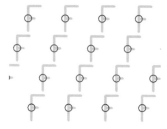

The translation grid (group 1)

Preparations
What happens if, in addition to translations, there are also reflections or glide reflections, but no rotations? Then all mirror or glide reflection axes must point in the same direction, because otherwise one could generate a rotation by multiplication.

We now assume that our discontinuous group $\mathcal{G}$ contains a *reflection*. If we conjugate $\mathcal{G}$ suitably, we can assume that a S_β is in $\mathcal{G}$ (because of Lemma 3.1.2). We will do that, because we are only interested in a characterization up to equivalence.

We start by choosing an $\mathbf{a}_0 \in \Delta$ with minimal norm. Then there are three possibilities:

1. It is $S_\beta \mathbf{a}_0 = \mathbf{a}_0$.
2. It is $S_\beta \mathbf{a}_0 = -\mathbf{a}_0$.
3. Neither 1. nor 2. applies.

Case 1: Suppose that $S_\beta \mathbf{a}_0 = \mathbf{a}_0$. We choose $\mathbf{b}_0 \in \Delta \setminus \mathbb{Z}\mathbf{a}_0$ with minimal norm. Then $\mathbf{a}_0, \mathbf{b}_0$ is a basis of Δ due to Lemma 3.3.4. Two cases are possible:

Case 1a: $\mathbf{b}_0$ *is perpendicular to* $\mathbf{a}_0$. Then $S_\beta \mathbf{b}_0 = -\mathbf{b}_0$. Consequently, all movements S_n, defined by $x \mapsto n\mathbf{b}_0 + S_\beta x$, are reflections in $\mathcal{G}$ and all reflections have this form. (If $S : x \mapsto s + S_\beta x$ is in $\mathcal{G}$, so is $S \circ S_\beta = T_\mathbf{s}$, i.e. $\mathbf{s} \in \Delta$. But s is – as $\mathbf{b}_0$ – orthogonal to the mirror axis so that $s = n\mathbf{b}_0$ for a suitable integer n since $\mathbf{a}_0\, \mathbf{b}_0$ is a basis of Δ.) In addition there are improper glide reflections $G_{n,m} :\ x \mapsto m\mathbf{a}_0 + n\mathbf{b}_0 + S_\beta x$.

Case 1b: $\mathbf{b}_0$ *is not perpendicular to* $\mathbf{a}_0$. If necessary, we can pass from $\mathbf{b}_0$ to $-\mathbf{b}_0$ and assume that the angle between $\mathbf{a}_0$ and $\mathbf{b}_0$ is less than $\pi/2$:

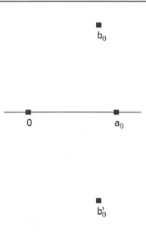

$\mathbf{a}_0$ and $\mathbf{b}_0$ enclose a sharp angle. Mirror axis: *gray*

(Attention: It will be shown immediately that a situation like in the picture cannot occur. But at the moment we don't know any better.) Necessarily are $\mathbf{b}_0$ and $\mathbf{b}_0' := S_\beta \mathbf{b}_0$ linear independent. (Note that $\mathbf{b}_0' \in \Delta$ because of proposition 2.3.5.) We claim that $\mathbf{b}_0 + \mathbf{b}_0' = \mathbf{a}_0$. In any case $\mathbf{b}_0 + \mathbf{b}_0'$ lies on the mirror axis, is therefore of the form $n\mathbf{a}_0$ for an $n \in \mathbb{N}$. ($n = 0$ and negative n do not come into question, since the enclosed angle is sharp.) Would $n \geq 2$, then $\mathbf{b}_0 - \mathbf{a}_0$ would have a smaller norm than $\mathbf{b}_0$: contradiction.

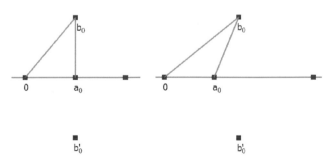

$n = 2, 3, \ldots$ lead to a contradiction. We see the cases $n = 2$ and $n = 3$. The vector $\mathbf{b} + \mathbf{b}_0$ is always *blue* marked

With $\mathbf{a}_0, \mathbf{b}_0$ is also $\mathbf{b}_0, \mathbf{b}_0'$ a basis of Δ.

Together: In case 1b one can choose a basis of Δ so that the mirror axis is the diagonal in a mesh of the translation grid.

Case 2: $\mathbf{a}_0$ *is perpendicular to the mirror axis.*[7] This means that $S_\beta \mathbf{a}_0 = -\mathbf{a}_0$. Choose $\mathbf{b}_0$ with minimal norm in $\Delta \setminus \mathbb{Z}\mathbf{a}_0$. Again, $\mathbf{a}_0, \mathbf{b}_0$ is a basis of Δ due to Lemma 3.3.4, and again there are two cases possible:

---

[7] W.l.o.g. the axis should be horizontal and $\mathbf{a}_0$ should point upwards.

Case 2a: $S_\beta \mathbf{b}_0 = \mathbf{b}_0$. $\mathbf{b}_0$ therefore lies on the mirror axis. This corresponds to Case 1a, with the roles of $\mathbf{a}_0$ and $\mathbf{b}_0$ reversed. In particular, the reflections are the images $\mathbf{x} \mapsto n\mathbf{a}_0 + S_\beta \mathbf{x}$ and the improper glide reflections are the motions $\mathbf{x} \mapsto m\mathbf{b}_0 + n\mathbf{a}_0 + S_\beta \mathbf{x}$.

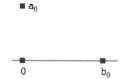

The image for Case 2a

Case 2b: $\mathbf{b}_0$ *does not lie on the mirror axis.* Possibly one has to go from $\mathbf{b}_0$ to $-\mathbf{b}_0$, $S_\beta \mathbf{b}_0$ or $-S_\beta \mathbf{b}_0$; then one may assume that $\mathbf{b}_0$ points to the right and up and $\mathbf{a}_0$ points up.

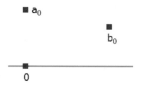

The image for case 2b (still preliminary!)

Let $\mathbf{b}_0' := S_\beta \mathbf{b}_0$. We claim: $\mathbf{b}_0 - \mathbf{b}_0' = \mathbf{a}_0$. $\mathbf{b}_0 - \mathbf{b}_0'$ is a vector in Δ that is orthogonal to the mirror axis, so it has the form $n\mathbf{a}_0$ for a $n \in \mathbb{N}$. And as in case 1b, one shows that $n \geq 2$ would lead to a contradiction, because then $\mathbf{b}_0 - \mathbf{a}_0$ would have a smaller norm than $\mathbf{b}_0$. So it must look like this

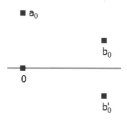

The typical image for case 2b: $\mathbf{b}_0$ and $\mathbf{b}_0'$ form a basis

If you choose $\mathbf{b}_0$ and $\mathbf{b}_0'$ as a basis, then the mirror axis is the diagonal of a rhombic translation grid.

Case 3: $\mathbf{a}_0$ *is oblique to the mirror axis.* $\mathbf{b}_0 := S_\beta \mathbf{a}_0$ has the same norm as $\mathbf{a}_0$, therefore (because of Lemma 3.3.4) $\mathbf{a}_0, \mathbf{b}_0$ is a basis of Δ. In this case too, the mirror axis is the diagonal of a rhombic translation grid.

This shows:

Proposition 3.5.5

Let $S_\beta \in \mathcal{G}$. Then two cases are possible:

(i)   You can choose a basis $\mathbf{a}_0, \mathbf{b}_0$ of Δ so that $S_\beta \mathbf{a}_0 = \mathbf{a}_0$ and $S_\beta \mathbf{b}_0 = -\mathbf{b}_0$.
     (The rectangular case.)

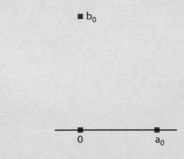

The typical rectangular case

Or:

(ii)   There is a basis $\mathbf{a}_0, \mathbf{b}_0$ of Δ so that $\mathbf{a}_0 + \mathbf{b}_0$ belongs to the mirror axis
      and $\mathbf{a}_0 - \mathbf{b}_0$ is perpendicular to it. (The rhombic case.)

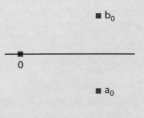

The typical rhombic case

Proposition 3.5.6

Let S_β be in $\mathcal{G}$ and suppose that there are no rotations in $\mathcal{G}$.

(i)   In the case of a rectangle, choose a basis $\mathbf{a}_0, \mathbf{b}_0$ of Δ as in the previous
     proposition. We claim:

     a) The translations in $\mathcal{G}$ are exactly the maps

$$T_{n,m} : \mathbf{x} \mapsto n\mathbf{a}_0 + m\mathbf{b}_0 \qquad (\text{with } n, m \in \mathbb{Z}).$$

     b) The reflections are exactly the maps

$$S_n : \mathbf{x} \mapsto n\mathbf{b}_0 + S_\beta \mathbf{x} \qquad (\text{with } n \in \mathbb{Z}).$$

c) The improper glide reflections are exactly the maps

$$G_{n,m} := T_{n,m} \circ S_\beta \qquad \text{(with } n, m \in \mathbb{Z}\text{)}.$$

d) There are no proper glide reflections.

(ii) In the rhombic case, choose a basis $\mathbf{a}_0, \mathbf{b}_0$ of Δ with $S_\beta(\mathbf{a}_0 + \mathbf{b}_0) = \mathbf{a}_0 + \mathbf{b}_0$ and $S_\beta(\mathbf{b}_0 - \mathbf{a}_0) = -(\mathbf{b}_0 - \mathbf{a}_0)$ according to proposition 3.5.5. The elements of $\mathcal{G}$ can then be described as follows:

a) The translations in $\mathcal{G}$ are exactly the mappings

$$T_{n,m} : \mathbf{x} \mapsto n\mathbf{a}_0 + m\mathbf{b}_0 \qquad \text{(with } n, m \in \mathbb{Z}\text{)}.$$

b) The reflections are exactly the mappings

$$S_n^\rho : \mathbf{x} \mapsto n(\mathbf{b}_0 - \mathbf{a}_0) + S_\beta \mathbf{x} \qquad \text{(with } n \in \mathbb{Z}\text{)}.$$

c) The improper glide reflections are exactly the mappings

$$G_{n,m}^\rho := \mathbf{x} \mapsto n(\mathbf{b}_0 - \mathbf{a}_0) + m(\mathbf{a}_0 + \mathbf{b}_0) + S_\beta \mathbf{x} \qquad \text{(with } n, m \in \mathbb{Z}\text{)}.$$

d) The proper glide reflections are exactly the mappings

$$\tilde{G}_{n,m}^\rho := \mathbf{x} \mapsto \mathbf{b}_0 + n(\mathbf{b}_0 - \mathbf{a}_0) + m(\mathbf{a}_0 + \mathbf{b}_0) + S_\beta \mathbf{x} \qquad \text{(with } n, m \in \mathbb{Z}\text{)}.$$

(iii) Let $\mathcal{G}$ and $\mathcal{G}'$ be discontinuous groups, both of which contain a reflection S_β resp. $S_{\beta'}$. If one of them is the rectangular case and the other is the rhombic case, they cannot be equivalent.

Proof (i) a) is clear because $\mathbf{a}_0, \mathbf{b}_0$ is a basis.

b) The $S_n = T_{n\mathbf{b}_0} \circ S_\beta$ lie in $\mathcal{G}$, and they are reflections, since $S_\beta(n\mathbf{b}_0) = -n\mathbf{b}_0$. Be the other way around S a reflection in $\mathcal{G}$. The mirror axis of S must be parallel to that of S_β, since otherwise a rotation in $\mathcal{G}$ would exist. S can therefore be written as $\mathbf{x} \mapsto sb + S_\beta \mathbf{x}$ with $S_\beta \mathbf{s} = -\mathbf{s}$. It is $S_\beta \circ S = T_\mathbf{s}$, i.e. $\mathbf{s} \in \Delta$. Therefore, $\mathbf{s} = t\mathbf{b}_0$ with a suitable $t \in \mathbb{R}$. Necessarily is t an integer n, because otherwise there would be a contradiction to the construction of $\mathbf{b}_0$. (Would be for example $\mathbf{s} = 2.7\mathbf{b}_0$, so $3\mathbf{b}_0 - \mathbf{s}$ would be an element from Δ, which is orthogonal to the mirror axis of S_β and has a smaller norm than $\mathbf{b}_0$.) Consequently, $S = S_n$.

c) This is clear, since the reflections are already characterized.

d) We assume that there is a proper glide reflection G in $\mathcal{G}$. The mirror axis must be parallel again to the one from S_β, i.e. $G : \mathbf{x} \mapsto \mathbf{s} + \mathbf{b} + S_\beta \mathbf{x}$, where $S_\beta \mathbf{s} = -\mathbf{s}$ and $S_\beta \mathbf{b} = \mathbf{b}$. The product $G \circ S_\beta$ is the translation $T_{\mathbf{s}+\mathbf{b}}$, so it is $\mathbf{s} + \mathbf{b} \in \Delta$. We write $\mathbf{s} + \mathbf{b} = n\mathbf{a}_0 + m\mathbf{b}_0$. Then it is necessary that $\mathbf{s} = n\mathbf{a}_0$ and $\mathbf{b} = m\mathbf{b}_0$, because $\mathbf{s}$ and $\mathbf{a}_0$ point in the direction of the mirror axis, and $\mathbf{b}$ and $\mathbf{b}_0$ are orthogonal to it. This shows that $G = G_{n,m}$ is not a proper glide reflection.

(ii) a) This is correct, since $\mathbf{a}_0, \mathbf{b}_0$ is a basis.

b) It is clear that all $S_n^\rho = T_{n(\mathbf{b}_0 - \mathbf{a}_0)} \circ S_\beta$ are reflections in $\mathcal{G}$, because $S_\beta n(\mathbf{b}_0 - \mathbf{a}_0) = -n(\mathbf{b}_0 - \mathbf{a}_0)$.

For the reversal, we first notice that between $\mathbf{0}$ and the vector $\mathbf{b}_0 - \mathbf{a}_0$ no element from Δ lies. If $\mathbf{b} := t(\mathbf{b}_0 - \mathbf{a}_0)$ were for a $t \in]0, 1[$ in Δ, then $\mathbf{b}_0 - \mathbf{b}$ would also be in Δ, and $\|\mathbf{b}_0 - \mathbf{b}\| < \|\mathbf{b}_0\|$ in contradiction to the construction of $\mathbf{b}_0$. This implies that the elements orthogonal to the mirror axis from Δ are just the $n(\mathbf{b}_0 - \mathbf{a}_0)$ with $n \in \mathbb{Z}$.

Let S now be a reflection in $\mathcal{G}$, this mapping can be written again as $\mathbf{s} + S_\beta \mathbf{x}$ with a S_β to the reflection axis orthogonal $\mathbf{s}$. Because of $S \circ S_\beta = T_\mathbf{s}$ we know that $\mathbf{s} \in \Delta$, and from the above observation it follows that $\mathbf{s} = n(\mathbf{b}_0 - \mathbf{a}_0)$ for a suitable $n \in \mathbb{Z}$. Thus $S = S_n^\rho$.

c) This is clear, since we already know the reflections.

d) We first claim that $G : \mathbf{x} \mapsto \mathbf{b}_0 + S_\beta \mathbf{x}$ is a proper glide reflection in $\mathcal{G}$. Certainly it is a glide reflection, but it is also proper. Otherwise one could write it as $S_{n,m}^\rho$. But then

$$\mathbf{b}_0 = n(\mathbf{b}_0 - \mathbf{a}_0) + m(\mathbf{a}_0 + \mathbf{b}_0),$$

and consequently $m - n = 0$ and $n + m = 1$. But that doesn't work for whole numbers n, m,

It has already been shown that all $\tilde{G}_{n,m}^\rho$ are proper glide reflections. Conversely, let $\tilde{G}$ be a proper glide reflection. The mirror axis is necessarily parallel to that of S_β, we write G as $G : \mathbf{x} \mapsto \mathbf{s} + \mathbf{b} + S_\beta \mathbf{x}$, where $S_\beta \mathbf{s} = -\mathbf{s}$ and $S_\beta \mathbf{b} = \mathbf{b}$. Because of $G \circ S_\beta = T_{\mathbf{s}+\mathbf{b}}$, $\mathbf{s} + \mathbf{b}$ lies in Δ, so $\mathbf{s} + \mathbf{b}$ can be represented as $n\mathbf{a}_0 + m\mathbf{b}_0$ with $n, m \in \mathbb{Z}$, and this vector has the form $n'(\mathbf{b}_0 - \mathbf{a}_0) + m'(\mathbf{a}_0 + \mathbf{b}_0) + \eta\mathbf{b}_0$, where $n', m' \in \mathbb{Z}$ and $\eta \in \{0, 1\}$. In the case $\eta = 0$, $G = G_{n',m'}^\rho$, this glide reflection would therefore not be proper, and in the case $\eta = 1$, $G = \tilde{G}_{n',m'}^\rho$ applies.

(iii) If $\mathcal{G}$ and $\mathcal{G}'$ are equivalent and $\mathcal{G}$ contains a proper glide reflection, then $\mathcal{G}'$ also contains a proper glide reflection. A glide reflection $G \in \mathcal{G}$ is only then improper if it can be written as the product of a translation $T_\mathbf{b}$ and a reflection S in $\mathcal{G}$ with $S \circ T_\mathbf{b} = T_\mathbf{b} \circ S$.

The claim now follows from the fact that one group contains proper glide reflections and the other does not. □

Group 2

The property There are, in addition to the translations, reflections and glide reflections.

The official notation W_1^1 or cm.

The analysis Based on the previous proposition, the rhombic case applies. We can choose a basis $\mathbf{a}_0, \mathbf{b}_0$ as in proposition 3.5.5 and then explicitly specify the translations, reflections, proper and improper glide reflections using proposition 3.5.6 (ii). The group table is easy to determine because we have the concrete representations. So, for example, $T_{n,m} \circ T_{n',m'} = T_{n+n',m+m'}$, and $S_n^\rho \circ S_{n'}^\rho = T_{n'-n,n-n'}$, etc. (Sometimes you have to be careful: For example, $T_{n,m} \circ S_k$ is a reflection or a proper or improper glide reflection, depending on whether $n - m$ is equal to zero or not equal to zero and even or not equal to zero and odd.)

Two groups of this type are equivalent If G and G' are such groups, we assume that both contain S_β (which can always be achieved by conjugation). Choose bases $\mathbf{a}_0, \mathbf{b}_0$ and $\mathbf{a}'_0, \mathbf{b}'_0$ as described above and represent the group elements explicitly. Then map the groups bijectively to each other: the translations by $T_{m\mathbf{a}_0+n\mathbf{b}_0} \mapsto T_{m\mathbf{a}'_0+n\mathbf{b}'_0}$, and analogously the reflections and glide reflections. Since, due to the construction, the group operations are the same in both groups, it is a group isomorphism, and since translations or reflections or … are mapped to translations or reflections or …, it is a group equivalence.

There is an example, and it is even a symmetry group

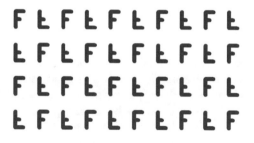

In the symmetry group are translations, glide reflections and proper glide reflections

Visualization

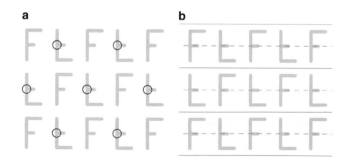

a Translation grid. **b**: Symmetry axes (Group 2)

Group 3

The property There are only translations and reflections.

The official notation $\mathcal{W}_1^2$ *or pm.*

The analysis This time we are in the rectangular case. After choosing a suitable basis of Δ we can explicitly describe all elements of the group with proposition 3.5.6 (i).

Two groups of this type are equivalent We proceed similarly as in the case of the previous group: identify movements, calculate group tables and use this description to define a mapping.

There is an example, and it is even a symmetry group

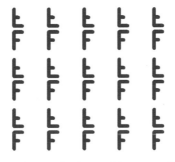

In the symmetry group are translations, reflections and improper glide reflections

Visualization

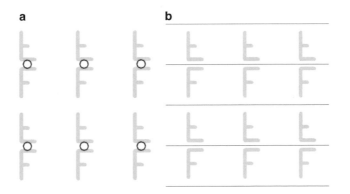

a Translation grid. **b** Symmetry axes (group 3)

Further preparations

We still have to consider the case that the discontinuous group $\mathcal{G}$ only contains proper glide reflections. To do this, we fix a proper glide reflection, and after conjugation of the group we can assume that it has the form $G : \mathbf{x} \mapsto \mathbf{v}_0 + S_\beta \mathbf{x}$, where $S_\beta \mathbf{v}_0 = \mathbf{v}_0$ and $\mathbf{v}_0 \neq \mathbf{0}$.

Let $\mathbf{a}_0$ be the shortest of the nonzero elements in Δ with the property $S_\beta \mathbf{a}_0 = \mathbf{a}_0$. The vector $2\mathbf{v}_0$ is a candidate, and after composition with a suitable translation we may assume that $2\mathbf{v}_0 = \mathbf{a}_0$ holds.

Choose further a $\mathbf{b}_0$ that has the smallest norm among the elements from $\Delta \setminus \mathbb{Z}\mathbf{a}_0$. We note that $S_\beta \mathbf{b}_0 \in \Delta$ (proposition 2.3.5).

The following cases have to be distinguished:

Case 1: $\mathbf{b}_0$ *is not perpendicular to* $\mathbf{a}_0$.

Case 2: $\mathbf{b}_0$ *is perpendicular to* $\mathbf{a}_0$. Here, two sub-cases will have to be considered:

Case 2a: $\mathbf{a}_0, \mathbf{b}_0$ *is a basis of* Δ.

Case 2b: $\mathbf{a}_0, \mathbf{b}_0$ *is not a basis of* Δ.

Proposition 3.5.7

Let $\mathcal{G}$ be a discontinuous group that contains no rotations and reflections, but proper glide reflections.

(i)   With the above notations, only Case 2a can occur.

(ii) The movements in $\mathcal{G}$ can then be described as follows:

a) The translations in $\mathcal{G}$ are exactly the maps

$$T_{n,m} : \mathbf{x} \mapsto n\mathbf{a}_0 + m\mathbf{b}_0 \qquad \text{(with } n, m \in \mathbb{Z}\text{)}.$$

b) The proper glide reflections are exactly the maps

$$\tilde{G}_{n,m} := \mathbf{x} \mapsto \mathbf{v}_0 + n\mathbf{a}_0 + m\mathbf{b}_0 + S_\beta \mathbf{x} \qquad \text{(with } n, m \in \mathbb{Z}\text{)}.$$

Proof (i) Assume we are in Case 1. Without restriction, $\mathbf{a}_0$ points to the right and $\mathbf{b}_0$ points to the right and up. Then it is necessary that $\mathbf{b}_0 + S_\beta \mathbf{b}_0 = \mathbf{a}_0$, because $\mathbf{b}_0 + S_\beta \mathbf{b}_0$ has the form $n\mathbf{a}_0$ for a $n \in \mathbb{N}$, and if one could subtract an $\mathbf{a}_0$, $\mathbf{b}_0$ would not have been minimal. $\mathbf{b}_0, S_\beta \mathbf{b}_0$ is a basis of Δ, because every $\mathbf{0}, \mathbf{b}_0, S_\beta \mathbf{b}_0, \mathbf{b}_0 + S_\beta \mathbf{b}_0$ different from $\mathbf{a} \in \Delta$ in the parallelogram spanned by $\mathbf{b}_0, S_\beta \mathbf{b}_0$ would imply a contradiction to the minimality of $\mathbf{b}_0$. $\mathbf{b}_0, \mathbf{b}_0^* := S_\beta \mathbf{b}_0$ are therefore a rhombic basis of Δ.

We could then find reflections in $\mathcal{G}$. From $\mathbf{b}_0 + \mathbf{b}_0^* = \mathbf{a}_0$ it follows that

$$S_\beta(\mathbf{b}_0 - \mathbf{v}_0) = S_\beta \left(\mathbf{b}_0 - \frac{\mathbf{b}_0 + \mathbf{b}_0^*}{2} \right) = -(\mathbf{b}_0 - \mathbf{v}_0),$$

i. e. $\mathbf{b}_0 - \mathbf{v}_0$ is perpendicular to the mirror axis. And that is why $T_{\mathbf{b}_0} \circ G^{-1} : \mathbf{x} \mapsto \mathbf{b}_0 - \mathbf{v}_0 + S_\beta \mathbf{x}$ is a reflection in $\mathcal{G}$. (G was introduced in the preparations.)

Case 2b cannot occur either. In this case, there must be an $\mathbf{a} \in \Delta$ different from the corners in the rectangle spanned by $\mathbf{a}_0, \mathbf{b}_0$. But this can only be the center of the rectangle, that is $c_0 := (a_0 + b_0)/2$: All others would imply a contradiction to the choice of a_0 and b_0.

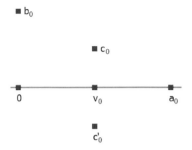

A rhombic basis in case 2b

Then the rhombic case applies again: $c_0, c_0' := S_\beta c_0$ is a basis of Δ. As was the case in the discussion of case 1, it would follow that there are reflections in $\mathcal{G}$.

(ii) We are therefore in Case 2a, and so part a) of the statement is clear. To prove b), we first note that all $\tilde{G}_{n,m}$ are proper glide reflections: This is because v_0 does not belong to Δ. Let G' be any glide reflection in $\mathcal{G}$. The mirror axis must point in the direction of the mirror axis of S_β, because otherwise there would be a non-trivial rotation. G' therefore has the form $G' : x \mapsto s + b + S_\beta x$, where $S_\beta s = -s$ and $S_\beta b = b$. $G' \circ G = T_{s+b+v_0}$, i.e. $s + b + v_0 \in \Delta$. Therefore, there are suitable $m, n \in \mathbb{Z}$ with $s + b + v_0 = n a_0 + m b_0$, thus $s + b = (n - 0{,}5)a_0 + m b_0$. Now s points in the direction of b_0 and b points in the direction of a_0, thus $s = m b_0$ and $b = (n - 0{,}5)a_0$. This shows that $G' = \tilde{G}_{n-1,m}$.

$\square$

Group 4

The property In addition to the translations, there are only proper glide reflections.

The official notation $\mathcal{W}_1^3$ or cg.

Two groups of this type are equivalent As a result of the foregoing investigations, all translations and the proper glide reflections can be explicitly specified after a canonically chosen basis has been selected. The group tables are identical, and thus one can define an equivalence between two groups of this type.

There is an example, and it is even a symmetry group

ㄴF ㄴ F ㄴ F ㄴ F

ㄴF ㄴ F ㄴ F ㄴ F

ㄴF ㄴ F ㄴ F ㄴ F

In the symmetry group there are only translations and proper glide reflections. The mirror axes go through the F-chains and parallel to them between the chains

Visualization

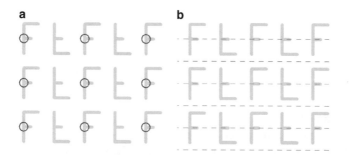

a Translation grid. **b** Symmetry axes (Group 4)

3.5.3   Translations, 2-Rotations, Reflections: 5 Groups

Group 5

The property There are only translations and 2-rotations.

The official notation W_2 or $p2$.

The analysis Since we are only interested in a characterization modulo equivalence, we may and will assume that $R_\pi \in \mathcal{G}$. Because of proposition 2.3.4 we have $\Delta_\pi = \Delta/2$, and thus all rotations and translations can be described explicitly:

- The translations are the $T_{n\mathbf{a}_0+m\mathbf{b}_0}$ with $m, n \in \mathbb{Z}$, where $\mathbf{a}_0, \mathbf{b}_0$ is a basis of Δ.
- The rotations are the $T_{n\mathbf{a}_0+m\mathbf{b}_0} \circ R_\alpha$ with $n, m \in \mathbb{Z}$. Note that if $\mathbf{x}_0$ is a 2-rotation center, then T has the form

$$\mathbf{x} \mapsto (\mathbf{x}_0 - R_\pi \mathbf{x}_0) + R_\pi \mathbf{x} = 2\mathbf{x}_0 + R_\pi \mathbf{x},$$

and the $2\mathbf{x}_0$ for $\mathbf{x}_0 \in \Delta/2$ are exactly the points from Δ.

Two groups of this type are equivalent If G and G' are presented with the above properties, conjugate so that R_π belongs to both. Then represent the translations and 2-rotations as above and map G bijectively on G'. The group operations remain, and translations or rotations pass into translations or rotations. The groups are therefore equivalent.

There is an example, and it is even a symmetry group.

In the symmetry group there are only translations and 2 rotations

Visualization As before, we represent the centers of 2-rotations by rhombi:

Translation grid and the rotation centers (group 5)

Preparations

We have to examine the interplay of 2-rotations and reflections in more detail. Given is a discontinuous movement group G, which contains β-reflections and 2-rotations.

The first conjugation As has been used several times before, we achieve by conjugation that S_β belongs to $\mathcal{G}$. There is also a rotation R by the angle π, we write it as $R : \mathbf{x} \mapsto \mathbf{a}' + R_\pi \mathbf{x}$. The vector $\mathbf{a}'$ we can decompose into two parts, one lies on the mirror axis, the other is orthogonal to it: $\mathbf{a}' = \mathbf{c} + \mathbf{d}$ with $S_\beta \mathbf{c} = -\mathbf{c}$ and $S_\beta \mathbf{d} = \mathbf{d}$.

The second conjugation We conjugate the group with $T_{-\mathbf{d}/2}$. This does not touch S_β, because $T_{-\mathbf{d}/2} \circ S_\beta \circ T_{\mathbf{d}/2}(\mathbf{x}) = S_\beta \mathbf{x}$. And from R we get

$$T_{-\mathbf{d}/2} \circ R \circ T_{\mathbf{d}/2}(\mathbf{x}) = -2\mathbf{d}/2 + \mathbf{a}' + \mathbb{R}_\pi x = \mathbf{c} + \mathbb{R}_\pi \mathbf{x}.$$

(In other words: W.l.o.g. we may assume that $\mathbf{a}'$ is orthogonal to the mirror axis.)

$R : \mathbf{x} \mapsto \mathbf{c} + R_\pi \mathbf{x}$ and S_β are therefore in $\mathcal{G}$. If you also note that $S_\beta \circ R \circ S_\beta$ is the rotation $R' : \mathbf{x} \mapsto -\mathbf{c} + R_\pi \mathbf{x}$ and that $R \circ R' = T_{2\mathbf{c}}$ is true, then $2\mathbf{c} \in \Delta$ follows. We summarize:

Lemma 3.5.8

A group $\mathcal{G}$ with the relevant properties here is conjugate to a group that contains S_β and $\mathbf{c} + R_\pi$, where $S_\beta \mathbf{c} = -\mathbf{c}$ and $2\mathbf{c} \in \Delta$ apply.

It is possible that not only $2\mathbf{c} \in \Delta$, but even $\mathbf{c} \in \Delta$ is true. Then $T_{-\mathbf{c}} \circ R = R_\pi$, and we are even allowed to assume that S_β *and* R_π are in $\mathcal{G}$.

Forget for the moment that there are rotations. Since S_β belongs to $\mathcal{G}$, we know from the previous section that there are two different possibilities:

- There is a basis $\mathbf{a}_0, \mathbf{b}_0$ of Δ with $S_\beta \mathbf{a}_0 = \mathbf{a}_0$ and $S_\beta \mathbf{b}_0 = -\mathbf{b}_0$.
  (The *rectangular case.*) Or:
- There is a basis $\mathbf{a}_0, \mathbf{b}_0$ of Δ with $S_\beta(\mathbf{a}_0 + \mathbf{b}_0) = \mathbf{a}_0 + \mathbf{b}_0$ and $S_\beta(\mathbf{a}_0 - \mathbf{b}_0) = -(\mathbf{a}_0 - \mathbf{b}_0)$. (The *rhombic case.*)

Let us first consider the rhombic case. $2\mathbf{c}$ belongs to Δ and is orthogonal to the mirror axis, so it can be written as $2\mathbf{c} = m(\mathbf{b}_0 - \mathbf{a}_0)$ for a suitable $m \in \mathbb{Z}$. If m is even, then $\mathbf{c}$ belongs to Δ, and we may assume that $R_\pi \in \mathcal{G}$ holds. If $m = 2k + 1$ is odd, then compose R with $T_{-k(\mathbf{b}_0 - \mathbf{a}_0)}$, i.e. we can even assume that $\mathbf{c} = (\mathbf{b}_0 - \mathbf{a}_0)/2$ is true.

In the following sketch, $\mathbf{c}$ is drawn for this case ($\mathbf{c} \notin \Delta$):

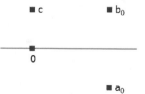

$\mathbf{c}$ is orthogonal to the mirror axis

In this case we apply a *third conjugation*, we conjugate with $T_{\mathbf{z}}$, where $\mathbf{z} := (\mathbf{a}_0 + \mathbf{b}_0)/4$. This results in S_β being converted into itself (because $S_\beta \mathbf{z} = \mathbf{z}$), and R becomes

$$\mathbf{x} \mapsto T_{-\mathbf{z}} \circ R \circ T_{\mathbf{z}} = \mathbf{c} - 2\mathbf{z} + R_\pi \mathbf{x} = -\mathbf{a}_0 + R_\pi \mathbf{x}.$$

It follows that $R_\pi = T_{\mathbf{a}_0} \circ T_{-\mathbf{z}} \circ R \circ T_{\mathbf{z}}$ belongs to $\mathcal{G}$.

If a reflection in $\mathcal{G}$ exists, then—modulo conjugation—three cases must be considered:

- *Case 1:* S_β, R_π in the rhombic case.
- *Case 2:* S_β, R_π in the rectangular case.
- *Case 3:* S_β and $\mathbf{c} + R_\pi$ in $\mathcal{G}$ in the rectangular case, where $S_\beta \mathbf{c} = -\mathbf{c}$ and $2\mathbf{c} \in \Delta, \mathbf{c} \notin \Delta$.

And what is known if *no* reflection belongs to $\mathcal{G}$, but (in addition to R_π-rotations) only glide reflections? We have already seen in Sect. 3.5.2 before the discussion of group 4 that the rectangular case must necessarily occur and that, after suitable conjugation, one can assume that there is a basis $\mathbf{a}_0, \mathbf{b}_0$ of Δ with the following properties: $S_\beta \mathbf{a}_0 = \mathbf{a}_0$, $S_\beta \mathbf{b}_0 = -\mathbf{b}_0$, and the proper glide reflections are described by $\mathbf{x} \mapsto \mathbf{a}_0/2 + \mathbf{a} + S_\beta \mathbf{x} (\mathbf{a} \in \Delta)$.

Let (with $\mathbf{v}_0 := \mathbf{a}_0/2$) G be the proper glide reflection $\mathbf{x} \mapsto \mathbf{v}_0 + S_\beta \mathbf{x}$ and $R : \mathbf{a}' + R_\pi$ a 2-rotation in $\mathcal{G}$. Write as just before $\mathbf{a}'$ as $\mathbf{c} + \mathbf{d}$ with $S_\beta \mathbf{c} = -\mathbf{c}$ and $S_\beta \mathbf{d} = \mathbf{d}$. Conjugate the group once more with $T_{-\mathbf{d}/2}$. Then

- $T_{-\mathbf{d}/2} \circ G \circ T_{\mathbf{d}/2} \mathbf{x} = G\mathbf{x}$ and
- $R'\mathbf{x} := T_{-\mathbf{d}/2} \circ R \circ T_{\mathbf{d}/2} \mathbf{x} = -2\mathbf{d}/2 + \mathbf{a}' + R_\pi \mathbf{x} = \mathbf{c} + R_\pi \mathbf{x}.$

Because of $S_\beta \mathbf{c} = -\mathbf{c}$ is $R''\mathbf{x} := G \circ R' \circ G^{-1}\mathbf{x} = -\mathbf{c} + R_\pi \mathbf{x}$ in $\mathcal{G}$, and $R' \circ R'' = T_{2\mathbf{c}}$. Again we have $2\mathbf{c} \in \Delta$.

Suppose it were even $\mathbf{c} \in \Delta$. Then would $R_\pi = T_{-\mathbf{c}} \circ R'$ be a group element and thus—because of $R_\alpha \circ S_\beta = S_{\beta+\alpha/2}$—also $R_\pi \circ G : \mathbf{x} \mapsto -\mathbf{v}_0 + S_{\beta+\pi/2}\mathbf{x}$. But that is a reflection, because $\mathbf{v}_0$ is perpendicular to the mirror axis of $S_{\beta+\pi/2}$. And there are no reflections according to the assumption. So $\mathbf{c}$ is of the form $(k + 0{,}5)\mathbf{b}_0$ for a $k \in \mathbb{Z}$, and after composition with $T_{-k\mathbf{b}_0}$ we may assume $\mathbf{c} = \mathbf{b}_0/2$. Now

$$R \circ G\mathbf{x} = (\mathbf{b}_0/2 + R_\pi) \circ (\mathbf{a}_0/2 + S_\beta)\mathbf{x} = (\mathbf{b}_0 + \mathbf{a}_0)/2 + S_{\beta+\pi/2},$$

and that is a proper glide reflection. It follows easily:

Proposition 3.5.9

If there are no reflections, but only glide reflections, then the rectangular case applies, and after choosing an appropriate basis $\mathbf{a}_0, \mathbf{b}_0$ of Δ, all glide reflections in $\mathcal{G}$ are of the form $\mathbf{x} \mapsto \mathbf{a}_0/2 + \mathbf{a} + S_\beta$ (glide reflection axis in direction β) or $\mathbf{x} \mapsto (\mathbf{b}_0 + \mathbf{a}_0)/2 + \mathbf{a} + S_{\beta+\pi/2}$ (glide reflection axis in direction $\beta + \pi/2$) for $\mathbf{a} \in \Delta$.

Proof It has already been shown that all $\mathbf{x} \mapsto \mathbf{a}_0/2 + S_\beta \mathbf{x}$ and all $\mathbf{x} \mapsto (\mathbf{a}_0 + \mathbf{b}_0)/2 + S_{\beta+\pi/2}\mathbf{x}$ are proper glide reflections. This means that all $\mathbf{x} \mapsto \mathbf{a} + \mathbf{a}_0/2 + S_\beta \mathbf{x}$ and all $\mathbf{x} \mapsto \mathbf{a} + (\mathbf{a}_0 + \mathbf{b}_0)/2 + S_{\beta+\pi/2}\mathbf{x}$ are proper glide reflections for $\mathbf{a} \in \Delta$. Let another proper glide reflection G be given. Only the directions β and $\beta + \pi/2$ can be used as mirror directions, because any other direction would lead to a rotation with a different angle of rotation than π. Let's say $G\mathbf{x} = \mathbf{a}' + S_\beta \mathbf{x}$. (The case $\beta + \pi/2$ can be treated analogously.) Write $\mathbf{a}' = \mathbf{c} + \mathbf{d}$ again, where $\mathbf{c}$ or $\mathbf{d}$ are multiples of $\mathbf{a}_0$ or $\mathbf{b}_0$. $G^2 = T_{2\mathbf{c}}$ then implies $2\mathbf{c} \in \Delta$, and as above, composition with a translation implies that G w.l.o.g. has the form $G\mathbf{x} = \mathbf{a}_0/2 + \mathbf{d} + S_\beta \mathbf{x}$. If $\mathbf{x} \mapsto -\mathbf{a}_0/2 + S_\beta \mathbf{x}$ is carried out first, then $\mathbf{d} \in \Delta$ follows, and thus G is already included in the catalog of known proper glide reflections. $\square$

In total, we therefore have four possible cases to discuss. You just have to characterize them by simple properties that are easy to find.

Group 6

The property There are translations, reflections in orthogonal directions and 2-rotations. Some of the rotation centers do not lie on a mirror axis.

The official notation $\mathcal{W}_2^1$ or *cmm*.

The analysis The above analysis shows that only the rhombic case is relevant.

Two groups of this type are equivalent This follows again from the above analysis. All occurring movements (2-rotations, translations, proper and improper glide reflections) can be described explicitly by choosing an appropriate basis of Δ. This then induces a group isomorphism between two groups with the same property, which is even an equivalence.

It should also be noted that there can only be axes of symmetry in the direction of β and $\beta + \pi/2$, otherwise a rotation by an angle different from π would be generated.

There is an example, and it is even a symmetry group

The symmetry group is the group 6

Visualization

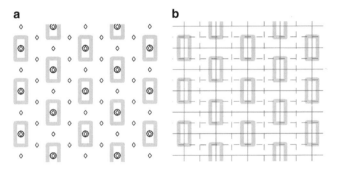

a Translation grid and the rotation centers. **b** Symmetry axes (Group 6)

Group 7

The Property There are translations, reflections in orthogonal directions and 2-rotations. All rotation centers lie on a mirror axis.

The official notation W_2^2 *or pm.*

The analysis Only one of the four possible cases meets this condition: It is the rectangular case with $c \in \Delta$. And again everything is explicitly describable.

Two groups of this type are equivalent This follows from the possibility of explicitly describing the group by suitable conjugation.

There is an example, and it is even a symmetry group

The symmetry group is group 7

Visualization

a   **b**

a Translation grid and the rotation centers. **b** Symmetry axes (group 7)

Group 8

The property There are only translations, 2-rotations, reflections, glide reflections; All mirror axes are parallel.

The official notation W_2^3 *or pmg.*

The analysis Now we are in the rectangular case with $c \notin \Delta$.

Two groups of this type are equivalent This results from the possibility of cataloging all occurring movements after choosing a suitable Δ-basis.

There is an example, and it is even a symmetry group

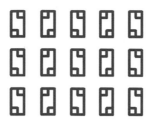

The symmetry group is group 8

Visualization

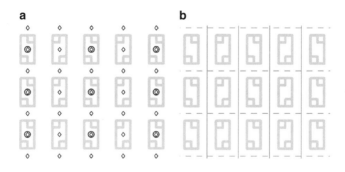

a Translation grid and the rotation centers. **b** Symmetry axes (group 8)

Group 9

The property There are only translations, 2-rotations and proper glide reflections.

The official notation W_2^4 or *pgg*.

The analysis The corresponding group corresponds to the last of the four cases discussed above. All rotations, translations and proper glide reflections are explicitly known.

Two groups of this type are equivalent Use the same strategy as above.

There is an example, and it is even a symmetry group

The symmetry group is group 9

Visualization

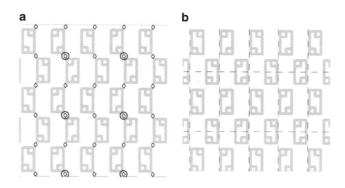

a Translation grid and the rotation centers. **b** Symmetry axes (group 9)

3.5.4   Translations, 3-Rotations, (Glide) Reflections: 3 Groups

Group 10

The property There are only translations and 3-rotations.

The official notation $\mathcal{W}_3$ or $p3$.

The analysis Let $\mathcal{G}$ be a discontinuous group with these properties. After conjugation, we may assume that $R_{2\pi/3}$ belongs to $\mathcal{G}$. This has an important consequence: Δ is a lattice of equilateral triangles. (Justification: Choose $\mathbf{a}_0$ in $\Delta \setminus \{\mathbf{0}\}$ with minimal norm. Set $\mathbf{b}_0 := R_{2\pi/3}\mathbf{a}_0$. Also $\mathbf{b}_0$ has minimal norm, and $\mathbf{a}_0, \mathbf{b}_0$ are linearly independent. Because of Lemma 3.3.4 then $\mathbf{a}_0, \mathbf{b}_0$ is a basis of Δ, whose meshes consist of rhombi with opening angles $60°$ and $120°$.)

In proposition 2.3.4 we have already characterized the rotation centers: They are the points from Δ together with the midpoints of the equilateral triangles emerging in the grid. (More precisely: If you have a grid of equilateral triangles and calculate the value $\mathbf{x}$ for any grid point $(\mathbf{x} - R_{-2\pi/3}\mathbf{x})/3$, you will obtain the altitude point of one of the triangles.)

From $R_{2\pi/3} \in \mathcal{G}$ it also follows that the rotations in $\mathcal{G}$ are exactly the movements $\mathbf{x} \mapsto \mathbf{a} + R_{\pm 2\pi/3}\mathbf{x}$: That they are all rotations is clear, and if R is a rotation in $\mathcal{G}$, then R is, according to the assumption, a rotation by $\pm 2\pi/3$, that is, of the form $R\mathbf{x} = \mathbf{a} + R_{\pm 2\pi/3}$. And because of $R \circ R_{\mp 2\pi/3} = T_\mathbf{a}, \mathbf{a} \in \Delta$ has to be true.

Two groups of this type are equivalent If $\mathcal{G}, \mathcal{G}'$ are groups of this type, then conjugate both first so that $R_{2\pi/3}$ belongs to it. Then choose bases $\mathbf{a}_0, \mathbf{b}_0$ or $\mathbf{a}_0', \mathbf{b}_0'$ of the translation lattice as described above. Then all translations and 3-rotations can be expressed using these bases: The translations or rotations in $\mathcal{G}$ are the $T_{n\mathbf{a}_0 + m\mathbf{b}_0}$ or the $T_{na_0+mb_0} \circ R_{\pm 2\pi/3}$ with $n, m \in \mathbb{Z}$. Finally, map $\mathcal{G}$ bijectively on $\mathcal{G}'$ using these representations: $T_{na_0+mb_0} \mapsto T_{na_0'+mb_0'}$, etc.

There is an example, and it is even a symmetry group

The symmetry group is the group 10

Visualization

The translation grid and the rotation centers (group 10)

Further preparations

What happens if $\mathcal{G}$ contains reflections and/or proper glide reflections in addition to 3-rotations? That there are only proper glide reflections is not possible: In the discussion of group 4 we saw that this can only happen in the case of rectangular translation grids.

Let $S \in \mathcal{G}$ be a reflection, $S : \mathbf{s} + S_\beta \mathbf{x}$. After a first conjugation of $\mathcal{G}$ we can assume that S_β belongs to $\mathcal{G}$. S_β leaves Δ invariant, and the Δ grid consists of isosceles triangles (see above). But there are only two directions that fulfill this condition:

Case 1 The line of symmetry runs through the sides of the triangle. If you imagine the β-direction horizontally, it looks like this:

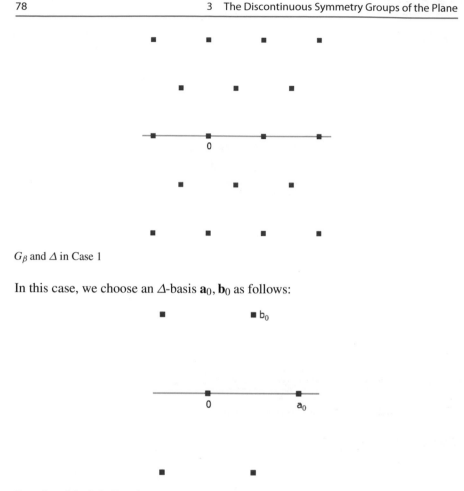

G_β and Δ in Case 1

In this case, we choose an Δ-basis $\mathbf{a}_0, \mathbf{b}_0$ as follows:

G_β and an Δ-basis in Case 1

Case 2 The line of symmetry intersects some of the triangle sides orthogonally in the middle. Again, we imagine the line of symmetry G_β horizontally in front of us, typically the situation is then as follows:

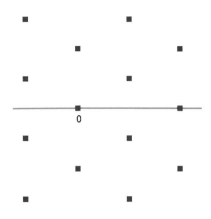

G_β and Δ in Case 2

In this case we will work with the following Δ basis:

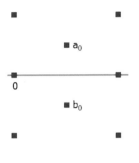

G_β and a Δ basis in Case 2

Proposition 3.5.10
After (possibly) another conjugation of $\mathcal{G}$ we are allowed to assume that $S_\beta, R_{2\pi/3} \in \mathcal{G}$.

Proof At the moment we only know that S_β and a 3-rotation $R : \mathbf{x} \mapsto \mathbf{a}' + R_{2\pi/3}\mathbf{x}$ belong to $\mathcal{G}$. Fix a non-zero $\mathbf{g}$ in the line of reflection G_β. We want to conjugate with $T_\mathbf{a}$, where $\mathbf{a} = t\mathbf{g}$ with $t \in \mathbb{R}$. Then $T_\mathbf{a} \circ S_\beta \circ T_{-\mathbf{a}} = S_\beta$ and $T_\mathbf{a} \circ R \circ T_{-\mathbf{a}}\mathbf{x} = \mathbf{a}' + t(\mathbf{g} - R_{2\pi/3}\mathbf{g}) + R_{2\pi/3}\mathbf{x}$. Since $\mathbf{g}$ and $\mathbf{g} - R_{2\pi/3}\mathbf{g}$ are linearly independent, we can choose $t \in \mathbb{R}$ so that $\mathbf{a}'' := \mathbf{a}' + t(\mathbf{g} - R_{2\pi/3}\mathbf{g})$ lies in G_β. If we conjugate with $T_{t\mathbf{a}}$ for this t, then S_β and the rotation $R' : \mathbf{x} \mapsto \mathbf{a}'' + R_{2\pi/3}\mathbf{x}$ with $\mathbf{a}'' \in G_\beta$ lie in $\mathcal{G}$.

With R' also $R'' := S_\beta \circ R' \circ S_\beta : \mathbf{x} \mapsto \mathbf{a}'' + R_{-2\pi/3}\mathbf{x}$ belongs to $\mathcal{G}$, because according to Lemma 2.2.4 (vi) we know that $S_\beta \circ R_{2\pi/3} \circ S_\beta = R_{-2\pi/3}$. It follows $R' \circ R''(\mathbf{x}) = \mathbf{a}'' + R_{2\pi/3}\mathbf{a}'' + \mathbf{x}$, i. e. $\mathbf{a}'' + R_{2\pi/3}\mathbf{a}'' \in \Delta$.

We claim that this implies $\mathbf{a}'' \in \Delta$, and then also $R_{2\pi/3} = T_{-\mathbf{a}''} \circ R'' \in \mathcal{G}$ would be as claimed. We discuss Case 1 and Case 2 separately:

Case 1: Write $\mathbf{a}'' = t\mathbf{a}_0$ for a suitable $t \in \mathbb{R}$: So all elements look like G_β. It is $R_{2\pi/3}\mathbf{a}_0 = \mathbf{b}_0 - \mathbf{a}_0$ and therefore

$$\mathbf{a}'' + R_{2\pi/3}\mathbf{a}'' = t\mathbf{a}_0 + t(\mathbf{b}_0 - \mathbf{a}_0) = t\mathbf{b}_0.$$

This vector belongs to Δ exactly when $t \in \mathbb{Z}$, it was thus $\mathbf{a}'' = t\mathbf{a}_0 \in \Delta$.

Case 2: This time we write $\mathbf{a}'' = t(\mathbf{a}_0 + \mathbf{b}_0)$. In the present case $R_{2\pi/3}\mathbf{b}_0 = \mathbf{a}_0 - \mathbf{b}_0$ and $R_{2\pi/3}\mathbf{a}_0 = -\mathbf{b}_0$. This implies

$$\mathbf{a}'' + R_{2\pi/3}\mathbf{a}'' = t(\mathbf{a}_0 + \mathbf{b}_0) + t\big((\mathbf{a}_0 - \mathbf{b}_0) - \mathbf{b}_0\big) = t(2\mathbf{a}_0 - \mathbf{b}_0).$$

That is an element of Δ exactly when $t \in \mathbb{Z}$. It follows again $\mathbf{a}'' \in \Delta$ and thus $R_{2\pi/3} \in \mathcal{G}$. $\square$

Group 11

The property There are translations and no other rotations except for 3-rotations. In addition, there are reflections and glide reflections, and all rotation centers lie on a mirror axis.

The official notation $\mathcal{W}_3^1$ or $p3m1$.

The analysis Because of the previous proposition, we can, for example, assume that $R_{2\pi/3}$ and S_β belong to $\mathcal{G}$. We already know the set $\Delta_{2\pi/3}$ of rotation centers, this set consists of Δ and the midpoints of the generated triangles.

Which mirror directions can there be? Certainly the directions β, $\beta + \pi/3$ and $\beta + 2\pi/3$, because these directions belong to S_β, $R_{2\pi/3} \circ S_\beta$ and $R_{2\pi/3}^2 \circ S_\beta$. Further directions are not possible, because then there would be rotations that are not 3-rotations. It follows that we are in Case 2, because in Case 1 not all rotation centers are on a mirror axis.

Let $S : \mathbf{x} \mapsto \mathbf{s} + S_\beta$ be a reflection in $\mathcal{G}$ with mirror axis direction β. Because of $S \circ S_\beta = T_\mathbf{s}$, $\mathbf{s} \in \Delta$, and $\mathbf{s}$ is perpendicular to G_β, it is therefore an integer multiple of $\mathbf{a}_0 - \mathbf{b}_0$. It follows: The mirror axes in direction β run parallel to G_β and go through the points $n(\mathbf{a}_0 - \mathbf{b}_0)/2$ for $n \in \mathbb{Z}$. Similarly the mirror axes in direction $\beta + \pi/3$ and $\beta + 2\pi/3$ can be described explicitly, and in this way one has a catalog of all reflections and not improper glide reflections.

What are the proper glide reflections in $\mathcal{G}$? We will first introduce a *strategy* here to identify them, which will also play an important role in the later sections. We proceed as follows:

The situation Given are a mirror direction β and an α -rotation. We assume that S_β and R_α belong to $\mathcal{G}$ and that we have already identified a basis $\mathbf{a}_0, \mathbf{b}_0$ of Δ. The problem: How do the glide reflections $G \in \mathcal{G}$ look in direction β?

The first step We write G as $G : \mathbf{x} \mapsto \mathbf{s} + \mathbf{b} + S_\beta \mathbf{x}$ (with $S_\beta \mathbf{s} = -\mathbf{s}$ and $S_\beta \mathbf{b} = \mathbf{b}$). We don't know much about $\mathbf{s}$ and $\mathbf{b}$ yet. But we know that $G^2 = T_{2\mathbf{b}}$, and therefore $2\mathbf{b} \in \Delta$ applies. $\mathbf{b}$ can therefore be "simply" expressed by $\mathbf{a}_0, \mathbf{b}_0$. And $\mathbf{s}$ too: One looks for the shortest vector $\mathbf{v}$ in Δ, which is orthogonal to the mirror axis, and can (possibly after transition from G to $T_{n\mathbf{v}} \circ G$ for a suitable $n \in \mathbb{Z}$) write $\mathbf{s}$ w.l.o.g. as $t\mathbf{v}$ with a $t \in [-0,5, 0,5]$.

The second step We consider $G \circ R_\alpha$, which is a reflection or glide reflection, the mirror axis runs in the direction of $\beta - \alpha/2$. For us, only the feed is interesting: This is the vector that arises when one projects $\mathbf{s} + \mathbf{b}$ orthogonally onto the new mirror axis. To calculate this, first determine the projections of $\mathbf{a}_0, \mathbf{b}_0$.

The final The projection of $\mathbf{s} + \mathbf{b}$ can now be explicitly given in the form $A(t)\mathbf{a}_0 + B(t)\mathbf{b}_0$ due to the preparations and the linearity of the projection, where A, B are simple functions of t. But $G \circ R_\alpha$ belongs to $\mathcal{G}$, the double of the feed must therefore be in Δ. This means $2A(t), 2B(t) \in \mathbb{Z}$, and from this the t can be easily determined.

In the present case we apply this strategy as follows. $G : \mathbf{x} \mapsto \mathbf{s} + \mathbf{b} + S_\beta \mathbf{x}$ (with $S_\beta \mathbf{s} = -\mathbf{s}$ and $S_\beta \mathbf{b} = \mathbf{b}$) is a proper glide reflection with mirror axis in the direction β in $\mathcal{G}$. For the feed $\mathbf{b}$, $2\mathbf{b} \in \Delta$ must hold, i.e. we may (possibly after subtracting a multiple of $\mathbf{a}_0 + \mathbf{b}_0$) assume that $\mathbf{b} = (\mathbf{a}_0 + \mathbf{b}_0)/2$. (Note that $\mathbf{b} \notin \Delta$, because the glide reflection should be proper.) And $\mathbf{s}$ is perpendicular to G_β, so it can be written as $t(\mathbf{a}_0 - \mathbf{b}_0)$, where (possibly after subtracting a multiple of $\mathbf{a}_0 - \mathbf{b}_0$) it may be assumed that $t \in [-0,5, 0, 5]$.

We now consider $G \circ R_{2\pi/3}$, which is a reflection or glide reflection with mirror axis in the direction $\beta - \pi/3$ (Lemma 2.2.4 (vi)). To decide what it is, we calculate the orthogonal projection P of $\mathbf{s} + \mathbf{b}$ onto $G_{\beta - \pi/3}$, which is the feed of $G \circ R_{2\pi/3}$. We have $P\mathbf{a}_0 = \mathbf{0}$ and $P\mathbf{b}_0 = \mathbf{b}_0 - \mathbf{a}_0/2$: To do this, you only need to use elementary properties of congruent triangles. So

$$P(\mathbf{s} + \mathbf{b}) = P\left(t(\mathbf{a}_0 - \mathbf{b}_0) + \frac{\mathbf{a}_0 + \mathbf{b}_0}{2} \right) = (0,5 - t)(\mathbf{b}_0 - \mathbf{a}_0/2).$$

The double of this feed vector, i.e. $(0,5 - t)(2\mathbf{b}_0 - \mathbf{a}_0)$, must belong to Δ. This is only possible for $t \in [-0,5, 0, 5]$ if $t = \pm 0,5$, that is, if $\mathbf{s} + \mathbf{b} \in \{\mathbf{a}_0, \mathbf{b}_0\}$.

Proper glide reflections in the direction β therefore have the form

$$\mathbf{x} \mapsto \mathbf{a}_0 + (n\mathbf{a}_0 + m\mathbf{b}_0) + S_\beta \mathbf{x}$$

or

$$\mathbf{x} \mapsto \mathbf{b}_0 + (n\mathbf{a}_0 + m\mathbf{b}_0) + S_\beta \mathbf{x}.$$

The mirror axes go through $(0,5 + m)\mathbf{a}_0$ $(m \in \mathbb{Z})$ and meet the points $(0,5 + n)\mathbf{b}_0$ $(n \in \mathbb{Z})$. Glide reflection axes in the direction $\beta + \pi/3$ and $\beta + 2\pi/3$ are treated analogously.

Two groups of this type are equivalent This follows as usual from the possibility to explicitly describe all the contained movements.

There is an example, and it is even a symmetry group

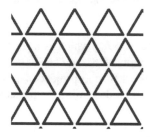

The symmetry group is group 11

Visualization

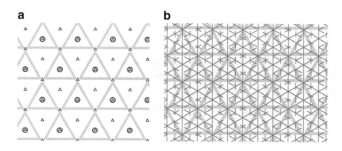

a Translation grid and the rotation centers. **b** Symmetry axes (group 11)

Group 12

The property There are translations and no other rotations except for 3-rotations. In addition, there are reflections and glide reflections, but not all rotation centers lie on a mirror axis.

The official notation W_3^2 or $p31m$.

The analysis This time case 1 is given. We already know the 3-rotations, and again there are only three mirror axis directions: β, $\beta + \pi/3$ and $\beta + 2\pi/3$. The vectors orthogonal to G_β in this case are the integer multiples of $2\mathbf{b}_0 - \mathbf{a}_0$, and thus all mirror axes and reflections are identified that belong to the direction β. Then all improper glide reflections are also known.

   To describe the proper glide reflections, we proceed similarly as with Group 11. If G is a proper glide reflection, then it has, for example, the form $\mathbf{x} \mapsto \mathbf{a}_0/2 + t(2\mathbf{b}_0 - \mathbf{a}_0)$ with a $t \in [-0{,}5, 0{,}5]$. This time, the projection P onto the mirror axis $G_{\beta-\pi/3}$ satisfies $P\mathbf{a}_0 = (\mathbf{a}_0 - \mathbf{b}_0)/2$, $P\mathbf{b}_0 = (\mathbf{b}_0 - \mathbf{a}_0)/2$. Consequently, the feed of $G \circ S_\beta$ is equal to

$$P\big(\mathbf{a}_0/2 + t(2\mathbf{b}_0 - \mathbf{a}_0)\big) = (1/4 - 3t/2)(\mathbf{a}_0 - \mathbf{b}_0).$$

And again, twice the feed, i.e. $(1/2 - 3t)(\mathbf{a}_0 - \mathbf{b}_0)$, must belong to Δ, and that is only possible for the $t \in [-0{,}5, 0{,}5]$ if $t = \pm 0{,}5$. The simplest proper glide reflections are therefore $\mathbf{x} \mapsto \mathbf{b}_0 + S_\beta \mathbf{x}$ and $\mathbf{x} \mapsto \mathbf{a}_0 - \mathbf{b}_0 + S_\beta \mathbf{x}$, and all others can be easily constructed from them (further feed $n\mathbf{a}_0$, further displacement orthogonal to G_β: $m(\mathbf{b}_0 - \mathbf{a}_0)$.) The mirror axes of proper glide reflections therefore go through the points $(0{,}25 + m/2)(\mathbf{a}_0 - \mathbf{b}_0)$ $(n, m \in \mathbb{Z})$.

Glide reflection axes in the direction of $\beta + \pi/3$ and $\beta + 2\pi/3$ are identified analogously.

Two groups of this type are equivalent If $\mathcal{G}$, $\mathcal{G}'$ have the properties required here, then the catalog of occurring movements induces a group equivalence.

There is an example, and it is even a symmetry group

The symmetry group is the group 12

Visualization

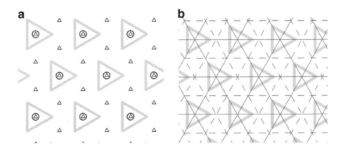

a Translation grid and the rotation centers. **b** Symmetry axes (group 12)

3.5.5   Translations, 4-Rotations, Reflections: 3 Groups

Group 13

The property There are translations and 4-rotations, but no reflections or glide reflections.

The official notation or *p4*.

The analysis There can only be 2- and 4-rotations (proposition 3.5.4), and we assume again—by conjugation—that $R_{2\pi/4} \in \mathcal{G}$. A basis of Δ is quickly found: Choose $\mathbf{a}_0$ with minimal positive norm in Δ and set $\mathbf{b}_0 := R_{2\pi/4}\mathbf{a}_0$ (cf. Lemma 3.3.4).

Then we know from proposition 2.3.4 Δ and all 2- and 4 -rotation centers: They are the points from $\Delta/2$. We can imagine it as sketched below in the visualization.

Everything (4-rotations, 2-rotations, translations) can be explicitly described: The translations are the $T_{m\mathbf{a}_0+n\mathbf{b}_0}$, and the rotations are exactly the conjugates of translations with powers of $R_{2\pi/4}$. The group table is easy to create.

Two groups of this type are equivalent This follows again from the fact that we have a catalog of all movements in $\mathcal{G}$, which only uses $\mathbf{a}_0, \mathbf{b}_0$.

There is an example, and it is even a symmetry group

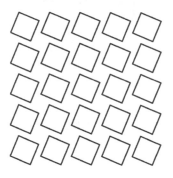

The symmetry group is the group 13

Visualization The following is the translation grid and the rotation centers are sketched. Diamonds are 2er-centers, small squares 4er-centers. (Note: 4er centers are also 2er-centers.)

The translation grid and the rotation centers for group 13

Further preparations

What can be said if $\mathcal{G}$ in addition to translations, 2- and 4-rotations contains further movements? There must be a reflection or a proper glide reflection. We choose a fixed translation mesh, which is spanned by $\mathbf{a}_0$ and $\mathbf{b}_0$. These vectors are orthogonal and have the same length.

Let G_β be a mirror or glide mirror axis that occurs. Since S_β leaves the set Δ invariant (proposition 2.3.5), there are only the following possibilities for the directions of possible symmetry axes[8]:

- The mirror axis runs parallel to a basis vector, for example parallel to $\mathbf{a}_0$.
- The mirror axis runs parallel to the angle bisector between the basis vectors.

Now we make use of the following fact: If S is a reflection or glide reflection in the direction β, then the $S \circ R_{2\pi/4}^k$ ($k = 0, 1, 2, 3$) are reflections or glide reflections in all directions $\beta + k\pi/4$. In each of the two cases, there is therefore a reflection direction that is parallel to $\mathbf{a}_0$. This means that we only have to investigate two cases systematically:

Case 1 G_β is parallel to $\mathbf{a}_0$, and G_β is the axis of reflection of a reflection. (We imagine G_β to be horizontal again.)

There is therefore in addition to $R_{2\pi/4}$ also a reflection $S : \mathbf{x} \mapsto \mathbf{s} + S_\beta \mathbf{x}$ in $\mathcal{G}$, where $S\mathbf{s} = -\mathbf{s}$. Since $\mathbf{s}$ is perpendicular to G_β, $\mathbf{s} = t\mathbf{b}_0$ applies to a $t \in \mathbb{R}$. To determine t, we use the strategy described in Sect. 3.5.4 under "Group 11". We consider $S \circ R_{2\pi/4}$. This is a glide reflection with reflection axis in the direction of $\beta - 2\pi/8$ (45° from upper left to lower right). For the projection P onto this reflection axis (the feed of the corresponding glide reflection), $P\mathbf{a}_0 = (\mathbf{a}_0 - \mathbf{b}_0)/2$ and $P\mathbf{b}_0 = (\mathbf{b}_0 - \mathbf{a}_0)/2$ apply. The feed is therefore $t(\mathbf{b}_0 - \mathbf{a}_0)/2$, and since twice the feed must belong to Δ, it follows that $t \in \mathbb{Z}$, i.e. $\mathbf{s} \in \Delta$. But then $S_\beta = T_{-\mathbf{s}} \circ S \in \mathcal{G}$ is also true.

---

[8] Note: This only applies to the directions. The actual mirror lines can be parallel-shifted to it.

It follows that the reflections (and thus the improper glide reflections) with reflection axis in the direction of β can be described explicitly: They are the $m\mathbf{a}_0 + n\mathbf{b}_0 + S_\beta$. The reflection axes go through $n\mathbf{b}_0/2$ for $n \in \mathbb{Z}$.

And what proper glide reflections in this direction, i.e. mappings of the form $G = \mathbf{s} + \mathbf{b} + S_\beta$ with $S_\beta \mathbf{s} = -\mathbf{s}$ and $S_\beta \mathbf{b} = \mathbf{b}$ are there? Twice the feed, i.e. $2\mathbf{b}$, must belong to Δ; and if the glide reflection is proper, $\mathbf{b} \notin \Delta$ must hold. So one has (possibly after combination with $T_{n\mathbf{a}_0}$ for a suitable $n \in \mathbb{Z}$) $\mathbf{b} = \mathbf{a}_0/2$. And $\mathbf{s}$ we can write as $t\mathbf{b}_0$, where w.l.o.g. $t \in [-0{,}5, 0{,}5]$ applies. As before, we go over to $G \circ R_{2\pi/4}$. The feed of this glide reflection is $P(\mathbf{s} + \mathbf{b}) = (1/2 - t)(\mathbf{a}_0 - \mathbf{b}_0)/2$. Twice that must be in Δ, and that is only possible for $t = \pm 1/2$, i.e. in the case $\mathbf{s} + \mathbf{b} = (\mathbf{a}_0 \pm \mathbf{b}_0)/2$.

Can there be reflections and proper glide reflections at the same time? Then S_β and $G : \mathbf{x} \mapsto (\mathbf{a}_0 \pm \mathbf{b}_0)/2 + S_\beta \mathbf{x}$ would be in $\mathcal{G}$, and thus also $G \circ S_\beta = T_{(\mathbf{a}_0 \pm \mathbf{b}_0)/2}$. $(\mathbf{a}_0 \pm \mathbf{b}_0)/2 \in \Delta$ would follow, a contradiction. In short: reflections and proper glide reflections cannot coexist in the case considered here.

Case 2 G_β is parallel to $\mathbf{a}_0$, and there is no reflection whose mirror axis is parallel to G_β. Then there must be a proper glide reflection, and we already know how to describe it explicitly.

Together this means: Either there are only reflections with axes parallel to $\mathbf{a}_0$ and no proper glide reflections, or only proper glide reflections.

Group 14

The property There are translations and 4-rotations. Parallel to the sides of the minimal translation square, there are only axes of reflections and no proper glide reflections.

The official notation W_4^1 or *p4m*.

The analysis We are in Case 1. All reflections with mirror axis parallel to $\mathbf{a}_0$ can be described explicitly, and by multiplication with $R_{2\pi/4}^k$ ($k = 1, 2, 3$) one obtains the reflections in the direction of the other possible axes: In the direction of $\beta + \pi/2$ there are again only reflections and no proper glide reflections, in the directions of $\beta \pm \pi/4$ there are reflections and both proper and improper glide reflections.

Two groups of this type are equivalent As before, this follows from the possibility to describe all elements of such groups explicitly after choosing a Δ basis.

There is an example, and it is even a symmetry group

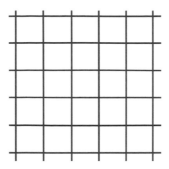

The symmetry group is group 14

Visualization

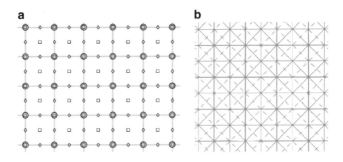

a Translation grid and the rotation centers. **b** Symmetry axes (group 14)

Group 15

The property There are translations and 4-rotations. Parallel to the sides of the minimal translation square, there are only axes of proper glide reflections.

The official notation W_4^2 or $p4g$.

The analysis This time we are in Case 2: only proper glide reflections with mirror axis parallel to $\mathbf{a}_0$. Again, everything is explicitly describable.

Two groups of this type are equivalent We argue here as before.

There is an example, and it is even a symmetry group

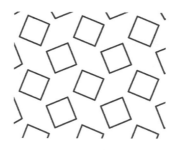

The symmetry group is group 15

Visualization

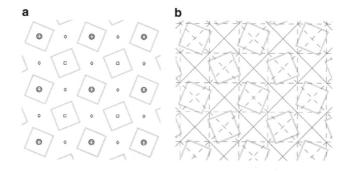

a Translation grid and the rotation centers. **b** Symmetry axes (group 15)

3.5.6   Translations, 6-Rotations, Reflections: 2 Groups

Group 16

The property There are only translations, 2-, 3- and 6-rotations.

The official notation $\mathcal{W}_6$ or *p6*.

The analysis One may, for example, assume that $R_{2\pi/6} \in \mathcal{G}$. Since then 2-rotations (namely $R_{2\pi/6}^3$) and 3-rotations ($R_{2\pi/6}^2$) also belong to $\mathcal{G}$, all movements and rotation centers can be read off from already known results.

Two groups of this type are equivalent With two groups having this property, one only needs to achieve, through conjugation, that $R_{2\pi/6}$ is included, and then choose bases of Δ, respectively, that enclose an angle of 60°. Then, there are identical catalogs of all movements in both groups, indexed by the bases.

There is an example, and it is even a symmetry group

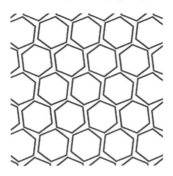

The symmetry group is group 16

Visualization You see the 6-, 3- and 2-rotation centers: hexagons, triangles, dia-monds. 3- and 2-centers, which are also 6-centers, were not drawn.

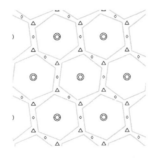

The translation grid and the rotation centers for group 16

Group 17

The property There are translations, 2-, 3- and 6-rotations as well as reflections.

The official notation W_6^1 or $p6m$.

The analysis It is tempting to use the results from Sects. 3.5.3 and 3.5.4, because with $R_{2\pi/6}$ there are also 2- and 3-rotations. But it is more convenient to apply the techniques used in the previous sections again.

So start with a basis $\mathbf{a}_0, \mathbf{b}_0$ of Δ, where $\mathbf{a}_0$ has minimal norm and $\mathbf{b}_0$ is obtained from $\mathbf{a}_0$ by a 60° rotation counterclockwise.

Now there is a mirror or glide mirror axis in the direction of β. Since Δ must be invariant under S_β, only few directions are possible, and rotations of the allowed mirror axes by 30° also lead to allowed mirror axes. (For this one has only to mul-tiply with $R_{2\pi/6}$.) And that means that an allowed mirror axis runs parallel to $\mathbf{a}_0$.

What do possible reflections look like then? Again we use the strategy described in the analysis of group 11. There must be movements of the type $S = t(2\mathbf{b}_0 - \mathbf{a}_0) + S_\beta$, where we can achieve $t \in [-0{,}5, 0{,}5]$. If we go to $S \circ R_{2\pi/6}$, this is a reflection in the direction of $\beta - \pi/6$, and the double of the feed lies exactly then in Δ, when $t = 0$. That means it was $S = S_\beta$, and consequently all reflections in this direction are given by $n(2\mathbf{b}_0 - \mathbf{a}_0) + S_\beta$ (with $n \in \mathbb{Z}$).

We now analyze proper glide reflections. They can be brought—with a t in the interval $[-0{,}5, 0{,}5]$—in the form $G = \mathbf{a}_0/2 + t(2\mathbf{b}_0 - \mathbf{a}_0) + S_\beta$. The double feed for $G \circ R_{2\pi/6}$ is exactly in Δ, if $t = \pm 1$, i.e., one has $S = \mathbf{a}_0 - \mathbf{b}_0 + S_\beta$ or $S = \mathbf{b}_0 + S_\beta$. Together this means:

- If a reflection is in $\mathcal{G}$, so are proper glide reflections, and all are known.
- If a proper glide reflection is in $\mathcal{G}$, so are reflections, and all are known.

You can therefore assume that $S_\beta \in \mathcal{G}$, and thus all reflections and glide reflections in the possible directions $\beta + k\pi/6$ $(k = 0, \ldots, 5)$ are explicitly describable.

Two groups of this type are equivalent This follows from the above considerations.

There is an example, and it is even a symmetry group

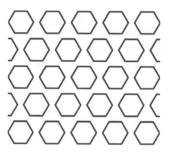

The symmetry group is group 17

Visualization

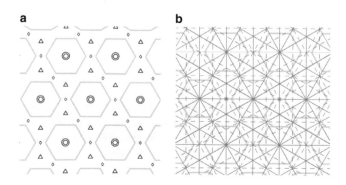

a Translation grid and the rotation centers. **b** Symmetry axes (group 17)

3.5.7   Classification: A Test

Below you will see 17 images, each symmetry group is one of the plane crystal groups. You, dear reader, are invited to identify the correct group for each image. The correct answers can be found at the end of this section.

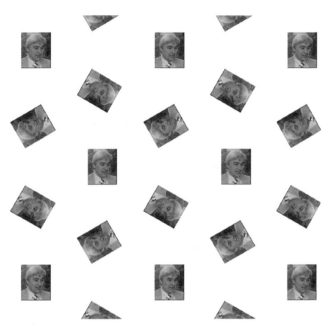

Example 1

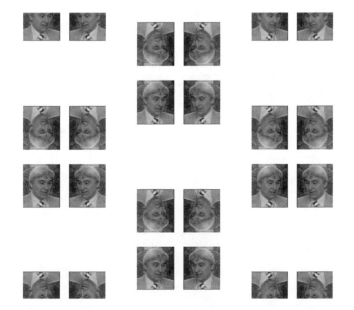

Example 2

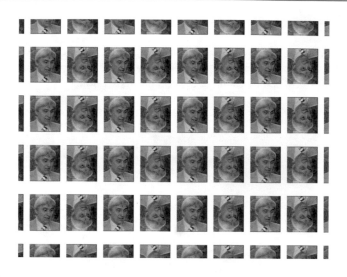

Example 3

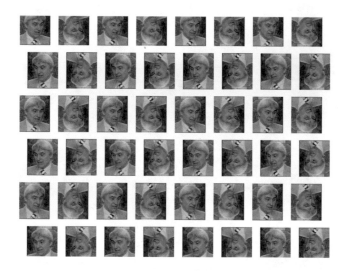

Example 4

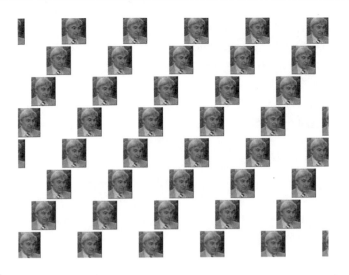

Example 5

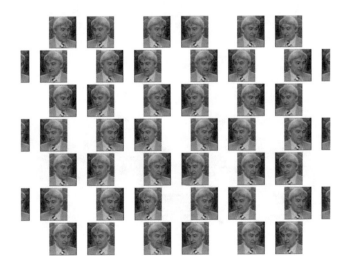

Example 6

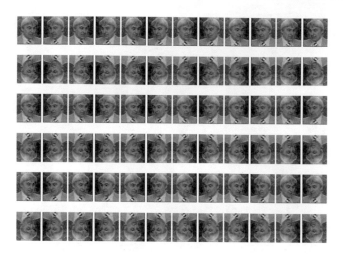

Example 7

Example 8

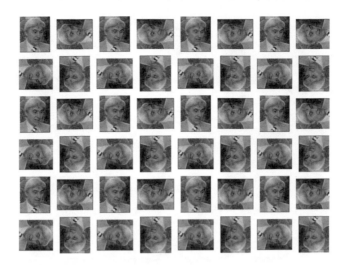

Example 9

Example 10

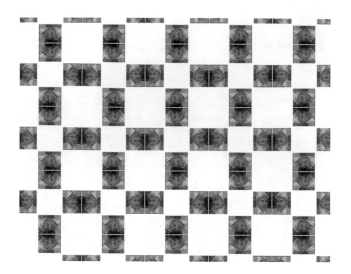

Example 11

Example 12

Example 13

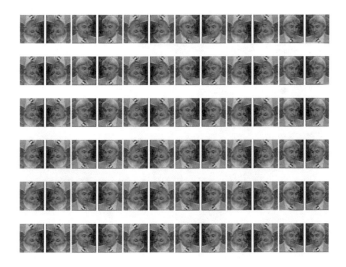

Example 14

Example 15

Example 16

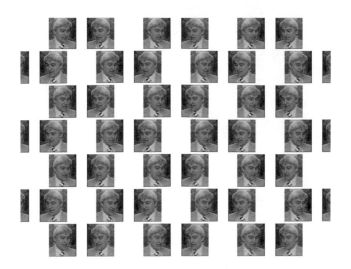

Example 17

Did you find the correct symmetry group? Compare:

Example	1	2	3	4	5	6	7	8	9
Group	$\mathcal{W}_3$	$\mathcal{W}_2^1$	$\mathcal{W}_1^3$	$\mathcal{W}_2^4$	$\mathcal{W}_1$	$\mathcal{W}_1^1$	$\mathcal{W}_2^2$	$\mathcal{W}_6^1$	$\mathcal{W}_4$

Example	10	11	12	13	14	15	16	17
Group	$\mathcal{W}_6$	$\mathcal{W}_6^2$	$\mathcal{W}_3^2$	$\mathcal{W}_4^1$	$\mathcal{W}_2^3$	$\mathcal{W}_3^1$	$\mathcal{W}_2$	$\mathcal{W}_1^2$

The Heesch Constructions

4

The German mathematician *Heinrich Heesch* (1906–1995) investigated how the above theoretical results could be implemented in practice. What does one have to do to design an image with a given symmetry group?

Heesch, who also made groundbreaking progress towards solving the four color theorem, gave a list of 28 construction possibilities for fundamental domains in the book "Flächenschluss" (from 1963, together with O. Kienzle). However, he approached the question as an engineering problem: What shapes can be cut out of a sheet of metal without trimming?

In this chapter we want to describe these Heesch constructions. We begin with important preparations in Sect. 4.1, which significantly reduces the number of situations that need to be investigated. Section 4.2 sketches how the constructions are obtained, we study some examples by way of illustration. And in Sect. 4.3 follows a systematic compilation of the methods.

In the recommended book "Heinrich Heesch" by Hans-Günther Bigalke you can learn a lot about the highs and lows of Heesch's life.

He spent his childhood and youth until his A-levels (1925) in Kiel. He was a good student, and it quickly became apparent that he was also musically gifted: He played the violin excellently. He then studied in Munich, his main subject was physics, in particular atomic physics. In parallel, he continued his musical education, becoming concertmaster for violin at a young age.

He then went to Zurich to do his doctorate. The topic of his dissertation came from crystallography, it was about a classification problem. So he had shifted his focus from physics to mathematics. He published his results and was recognized as a gifted young scientist in the mathematical world. In 1930 he went as an assistant to Hermann Weyl to Göttingen, the then Mecca of mathematics in Germany.

Actually, the next step of an academic standard career should have been taken now, the habilitation. But meanwhile, the National Socialists had taken power, and this led to dramatic changes at the universities. Many scientists (including

© The Author(s), under exclusive license to Springer Fachmedien Wiesbaden GmbH, part of Springer Nature 2022
E. Behrends, *Tilings of the Plane*, Mathematics Study Resources 2,
https://doi.org/10.1007/978-3-658-38810-2_4

Hermann Weyl) left Germany, and prospective habilitants had to complete several weeks of military training and ideological education. Heesch's habilitation was not recommended, and he left the university.

His economic situation was critical for many years, for a long time he had to be financially supported by his parents. But his special knowledge about crystallography and tilings was to work out very well for him. The interpretation of a tiling as the possibility of lossless punching of sheet metal was considered to be of war importance. He did not have to go to the front, but rather worked as a consultant for industrial companies.

In the post-war period he was then a "private scholar" in Kiel, where he also worked as a teacher at a grammar school for a while. There were two focuses in his scientific life. On the one hand, he continued to pursue the tiling problems. He also habilitated about this in 1957 and was able to gain a foothold as a private lecturer at the University of Hanover.

However, most of his scientific efforts in these years were devoted to the solution of the four-color problem: Can one color any map with four colors so that neighboring countries are colored differently? He develops most of the methods with which the four-color problem was then solved by Appel and Haken in 1977. Computers played a significant role in this proof, and Heesch was about to achieve a final solution. But he could not convince the German Research Foundation to finance the necessary computing time.

The chapter begins with the study of lattices and nets, they enable something like a combinatorial justification of the "construction plans" to be described later. We briefly motivate how construction specifications can be derived from nets, then the 28 Heesch constructions are explained and illustrated by examples.

4.1   Lattices and Nets

In the previous Chap. 3 the set Δ of $\mathbf{a} \in \mathbb{R}^2$ with $T_\mathbf{a} \in \mathcal{G}$ played an important role. If $\mathcal{G}$ is the symmetry group of an image F, then it is obviously enough to know the section of F in a translation mesh to be able to reconstruct F. Most of the time, even much less is enough, if $\mathcal{G}$ contains not only translations.

We have seen that there are not so many fundamentally different Δ. Depending on the group, the following can occur:

- the general lattice $\mathbb{Z}\mathbf{a} + \mathbb{Z}\mathbf{b}$;
- the rhombic lattice (the meshes are rhombic);
- the rectangular lattice;
- the square lattice;
- the hexagonal lattice (the meshes are equilateral triangles).

Here they are again:

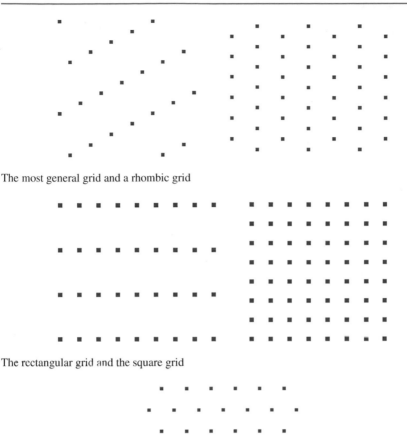

The most general grid and a rhombic grid

The rectangular grid and the square grid

The hexagonal grid

Then one knows, for example: If the image has 6-rotations as symmetries, then the translation grid must be hexagonal. Of course, this information is not enough to construct very general fundamental domains. We do not want to allow too pathological examples here, after all, it is about practically implementable results[1]. Therefore, we agree:

---

[1] Remember: The goal of Heesch/Kienzle was to stamp workpieces losslessly from sheet metal.

We want to construct bounded $F \subset \mathbb{R}^2$ that satisfy the following conditions:

- F is closed, simply connected, and F is the closure of the interior.
- The boundary of F is a piecewise smooth curve.
- F is a fundamental domain for one of the 17 plane crystallographic groups.

In the end, a method for the construction of such F should be given for a given symmetry group.

Assume we have such an F. By the action of the group operations on F, the plane is filled without gaps, and overlaps only occur at the edges. This has the following consequence: The boundary of F can be written as a sequence of consecutive curve sections $\gamma_0, \ldots, \gamma_{r-1}$, with the following properties:

- For each γ_i there is a γ_j, so that γ_i is converted by a suitable $T \in \mathcal{G}$ with $T \neq \mathrm{Id}$ to γ_j.
- If the endpoints of γ_i are denoted by P_i, P_{i+1} (so that $P_0 = P_r$), then at the P_i several copies of F meet (there are at least 3). The fundamental domain F therefore gives rise to a sequence of $v_0, \ldots, v_{r-1}$: v_i stands for the number of F-copies that meet at P_i. If only the edges of the F copies are considered, then v_i is the number of "edges" that meet at P_i.

As an example, let us consider an infinitely extended square grid, then all $T_{\mathbf{a}}$ with $\mathbf{a} \in \mathbb{Z}^2$ are symmetries. One can choose a square F as the fundamental domain, the four sides are the $\gamma_0, \ldots, \gamma_3$, and the sequence of multiplicities is $4, 4, 4, 4$. This "graph-theoretical condensate" of the tiling of the plane by copies of F is called a *lattice*. Precisely:

Definition 4.1.1

A *lattice* is an infinite graph in the plane, in which the meshes are generated by movements. If you run through the corners (clockwise or counterclockwise), a lattice gives rise to a finite sequence $v_0, \ldots, v_{r-1}$: The number r is the number of corners in a mesh, and $v_i \in \{3, 4, \ldots\}$ indicates how many edges meet in the i-th corner. (Since you can start at any point of the mesh to list the v_i, and since you can run through the corners clockwise or counterclockwise $v_0, \ldots, v_{r-1}$ is only uniquely determined up to cyclic permutation and reversal of the order.)

As examples, we consider a $(6, 6, 6)$-lattice and a $(4, 3, 4, 3, 3)$-lattice:

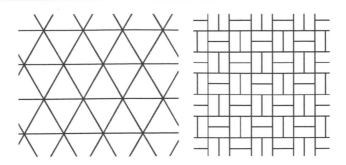

A $(6,6,6)$-lattice and a $(4,3,4,3,3)$-lattice

First of all, one has to understand which lattices can occur at all. Surprisingly, there are only a few possibilities for the numbers $v_0, \ldots, v_{r-1}$:

Theorem 4.1.2 (Laves)
One has

$$\sum_{i=0}^{r-1} \frac{1}{v_i} = \frac{r}{2} - 1.$$

Proof The proof is based on Euler's polyhedral theorem: If a subset is given in the plane by a surrounding line segment and if one divides the surface into finitely many parts by means of lines, then the following is true: (number of vertices) minus (number of edges) plus (number of parts) equals one. If we abbreviate the relevant numbers by E, K, F, then $E - K + F = 1$.

Now let us imagine that we have been given a net. We consider a gigantic section consisting of N meshes, which is bounded by an almost circular line segment. Then, obviously $F = N$. Each edge (with the exception of the edges on the border) is bordered by two surfaces, and each surface has r edges. In good approximation, K may be replaced by $rN/2$.

And how many corners are there? At each i-th corner of a mesh, v_i surfaces meet, so there are N/v_i corners of this type, and the total number is $N(1/v_0 + 1/v_1 + \cdots + 1/v_{r-1})$. We've simplified things a bit here, because this only applies to the corners inside our net cutout. Together, due to the Euler formula we conclude

$$N \sum_{i=0}^{r-1} \frac{1}{v_i} - \frac{Nr}{2} + N \approx E - K + F = 1.$$

After dividing by N, $\sum_{i=0}^{r-1} \frac{1}{v_i} - \frac{r}{2} + 1$ is therefore arbitrarily close to 0, which proves the claim.

However, the proof was simplified a bit. In a more detailed analysis, one would have to take into account that one can arrange it so that the order of magnitude of the number of boundary meshes is equal to $\sqrt{N}$. (Think of a $M \times M$ -square of square meshes. There $N = M^2$, and there are $4M - 4$ boundary meshes.) □

Lave's theorem provides the opportunity to systematically find all candidates. An immediate consequence is that r can be at most equal to 6: After all, $v_i \geq 3$, so the sum on the left can be at most equal to $r/3$:

$$\frac{r}{2} - 1 = \sum_{i=0}^{r-1} \frac{1}{v_i} \leq \frac{r}{3},$$

and that implies $r \leq 6$. And then, with the help of a computer, all tuples $v_0, \ldots, v_{r-1}$ with $r \leq 6$ and $v_0 \geq v_1 \leq \cdots \geq v_{r-1}$ can be found that fulfill this condition. It turns out:

- In the case of $r = 3$, possible are:

$$(6,6,6) \quad (8,8,4) \quad (10,5,5) \quad (12,6,4) \quad (12,12,3)$$
$$(15,10,3) \quad (18,9,3) \quad (20,5,4) \quad (24,8,3) \quad (42,7,3)$$

- In the case of $r = 4$, possible are:

$$(4,4,4,4) \quad (6,4,3,4) \quad (6,4,4,3) \quad (6,6,3,3) \quad (12,4,3,3)$$

- In the case of $r = 5$, possible are: $(4,4,3,3,3)$ and $(6,3,3,3,3)$.
- In the case of $r = 6$, only $(3,3,3,3,3,3)$ is possible.

To find specific networks for these numbers, of course, permutations of the above tuples must also be taken into account, with $(6,6,3,3)$ also $(6,3,3,6)$, $(3,6,3,6)$, etc. But which of these ordered tuples can be realized by networks? For those for which it is not possible, a proof must be given, and for the others an example must be provided. We show exemplarily:

Proposition 4.1.3
There is no $(6,6,3,3)$-net.

Proof Assume that one would find a $(6,6,3,3)$-net. We draw an arbitrary mesh F (see the following figure) and consider the mesh F' that is adjacent to the 6-6-edge.

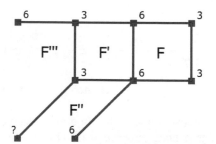

There is no $(6,6,3,3)$-net

Necessarily, there is a 3-3-edge, but how can it continue downwards and to the left?

The mesh adjacent to F' at the lower 3-6-edge would have to have a F''-branching after the 6 in a clockwise direction, and where the "?" is, there should be a 3.

But that can't be. If, for example, the mesh adjacent to F' at the 3-3-edge is called F''', then only the branching numbers 6 and 6 remain for the remaining corners of F'''. But at the corner with the "?" a 6 and a 3 can't stand at the same time.

□

And the positive? There are (with the exception of cyclic permutation and inversion of the v_i) exactly 11 nets, namely:

For $r = 3$ the nets $(6, 6, 6)$, $(12, 6, 4)$, $(12, 12, 3)$, $(8, 8, 4)$.

For $r = 4$ the nets $(4, 4, 4, 4)$, $(6, 3, 6, 3)$, $(6, 4, 3, 4)$.

For $r = 5$ the nets $(4, 3, 4, 3, 3)$, $(4, 4, 3, 3, 3)$, $(6, 3, 3, 3, 3)$.

For $r = 6$ the net $(3, 3, 3, 3, 3, 3)$.

Here are examples (we already know two):

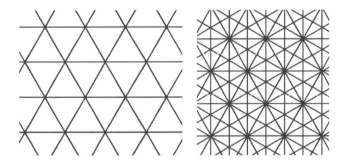

The nets $(6, 6, 6)$ and $(12, 6, 4)$

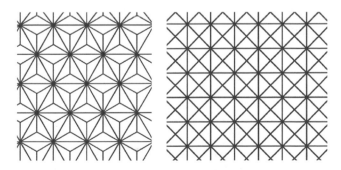

The nets $(12, 12, 3)$ and $(8, 8, 4)$

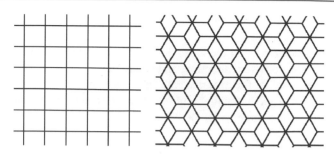

The nets $(4, 4, 4, 4)$ and $(6, 3, 6, 3)$

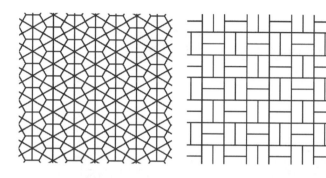

The nets $(6, 4, 3, 4)$ and $(4, 3, 4, 3, 3)$

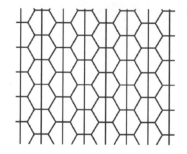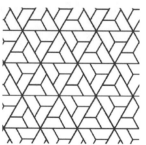

The nets $(4, 4, 3, 3, 3)$ and $(6, 3, 3, 3, 3)$

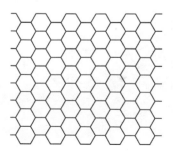

The net $(3, 3, 3, 3, 3, 3)$

4.2   The Heesch Construction: Motivation

Using the results of the previous section, fundamental domains are to be constructed now. After all, one knows:

- If F is a fundamental domain, then the $T(F)$ with $T \in \mathcal{G}$ give rise to a net. So if one writes the boundary pieces of F as $\gamma_0, \ldots, \gamma_{r-1}$, then $3 \le r \le 6$, and for each γ_i there is a group operation T and a γ_j with $T(\gamma_i) = \gamma_j$.
- There are two independent directions of translation, so that the plane is filled without gaps by translations of a set F' in these directions. Here, F' is the union of certain $T(F)$.

Suppose a γ_i goes through a translation $T_\mathbf{a}$ into a γ_j. Then γ_i and γ_j cannot be directly adjacent on the boundary of F. On the other hand, in many cases there are no further restrictions, one defines:

> **Definition 4.2.1**
> Let P_1, P_2 be different points of the plane. A *free line* from P_1 to P_2 is then a continuous non-intersecting curve from P_1 to P_2.

Here we only want to allow curves that are not too pathological: We will only consider lines that could be relevant in some artistic or technical context, and it is enough if we only allow continuous piecewise smooth curves.

And what needs to be considered when a π-rotation is applied to γ_i? Then γ_i must rotate by 180° about the center. This leads to

> **Definition 4.2.2**
> Let P_1, P_2 be different points of the plane. A *C-line*[2] from P_1 to P_2 is then a free line that additionally has the following property: If you rotate it around $(P_1 + P_2)/2$ by the angle π, it goes into itself.

Here are some examples of C-lines:

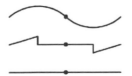

Examples of C-lines

---

[2] The "C" is meant to remind us of "center".

And now it can start. As an introductory example, we assume that $\mathcal{G}$ only contains translations and that the fundamental domain generates the net $(4, 4, 4, 4)$. Then F will consist of four edge pieces $\gamma_0, \gamma_1, \gamma_2, \gamma_3$, and γ_0 (or γ_1) will be transformed into γ_1 (or γ_2) by a translation.

In other words: You can choose four points freely that span a parallelogram and then you can choose two free lines. By moving, the fundamental domain of a group is generated, which consists only of translations. Here is an example:

Two free lines generate a fundamental domain

If you let the translations act, the following tiling results[3]:

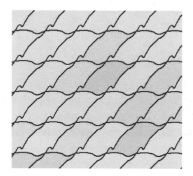

The tiling it produces

This is already the first Heesch construction, it is called *TTTT*. The "*T*" stands for Translation, in general one goes through the edges of the fundamental domain and describes by a letter which group operation acts on it. The following abbreviations are used:

"*T*"   for "Translation";
"*C*"   for "Rotation by 180°, that is a 2-rotation";
"*C*₃"   for "3-rotation";
"*C*₄"   for "4-rotation";

---

[3]To make it look more interestingly, the copies of the fundamental domain are colored differently here and in the following examples.

"C_6"   for "6-rotation";
"G"   for "glide reflection".

For another example, let's assume that we are talking about $(6, 6, 6)$ network. F consists of three edge pieces, and these can all be constructed as C-lines between three freely chosen points. A possible fundamental domain could then look like this[4]:

Three C-lines generate a fundamental domain

It will be convenient to apply as many of the allowed group operations as necessary to create an F' that has only to be moved to fill the plane. Such an F' is what we will call a *translation mesh*. In this case, we only need to rotate F at any center of rotation of a C-line. The resulting translation mesh and tiling are shown below (the translation vectors are also shown):

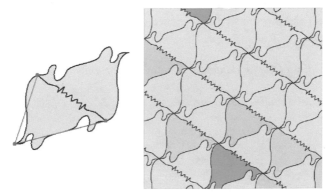

The translation mesh and tiling

That was Heesch construction number 3, abbreviated as CCC.

Reflections do not occur in Heesch constructions: Mirror symmetries must be enabled by special choices of the occurring free lines and C-lines.

---

[4] The centers of the C-lines are marked in red.

Heesch then systematically examined which letters from the set $\{T, C, C_3, C_4, C_6, G\}$ can be used for the γ_i in $\gamma_0, \ldots, \gamma_{r-1}$ if one knows that one of the 11 networks found above is the underlying one. We cannot reproduce this here, the derivation would be very extensive and would require many further preparations.

4.3   The Heesch Constructions: 28 Methods

The Heesch constructions are now described systematically, we follow the presentation in the book by Heesch-Kienzle.

The Heesch construction 01

Name TTTT.

The underlying network $(4, 4, 4, 4)$.

What can be given in advance? Two free lines.

Construction description Choose points A and B arbitrarily and then a free line between A and B (see the next figure). Move them to a line from C to D, the displacement vector is arbitrary. Choose a second free line from A to C (the left boundary line) and move it so that it connects B with D. This completes the fundamental domain.

Number of copies of the fundamental domain in the translation mesh One.

The fundamental domain and the translation mesh

Heesch 01: Fundamental domain and translation mesh

The associated tiling

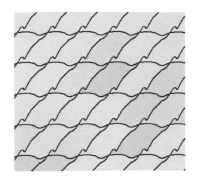

Heesch 01: The tiling

The symmetry group In the general case, the symmetry group is the *p*1; for notation see Sect. 3.5. With a special choice of the free lines (special angles between
the edges, symmetrical lines), larger groups can occur as symmetry groups here
and in the following examples.

Heesch construction 02

Name TTTTTT.

The underlying network $(3, 3, 3, 3, 3, 3)$.

What can you specify? Three free lines.

Construction description Choose *A* and *B* arbitrarily and then a free line between
A and *B*. Move it to a line from *C* to *D*, the displacement vector is arbitrary. Then
choose—with an arbitrary point *E*—a free line from *A* to *C* and move it so that it
connects *D* and *F*. (*E* therefore moves to *D* and *A* to *F*). Now connect with a third
free line *E* and *C*, which then has to be moved to a connecting line from *B* to *F*.

Number of copies of the fundamental domain in the translation mesh One.

The fundamental domain and the translation scheme

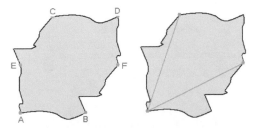

Heesch 02: Fundamental domain and translation scheme

The corresponding tiling

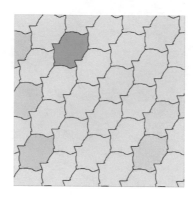

Heesch 02: The tiling

The symmetry group In general, the symmetry group is the $p1$.

The Heesch construction 03

Name CCC.

The underlying network $(6, 6, 6)$.

What can be specified? Three C-lines.

Construction description We draw three arbitrary points A, B and C in the plane and connect A and B, B and C, C and A each by an C-line.

Number of copies of the fundamental domain in the translation scheme Two.

The fundamental domain and the translation scheme

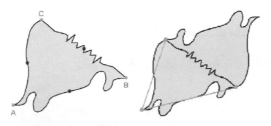

Heesch 03: Fundamental domain and translation scheme

The corresponding tiling

Heesch 03: The tiling

The symmetry group In the general case, the symmetry group is the *p*2.

The Heesch construction 04

Name CCCC.

The underlying network (4, 4, 4, 4).

What can be specified? Four C-lines.

Construction description This time we draw four points *A*, *B*, *C* and *D* in the plane and connect *A* and *B*, *B* and *C*, *C* and *D*, *D* and *A* each by a *C*-line.

Number of copies of the fundamental domain in the translation scheme Four.

The fundamental domain and the translation scheme

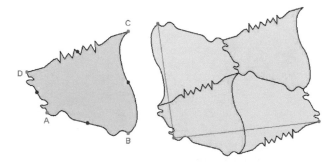

Heesch 04: Fundamental domain and translation scheme

The associated tiling

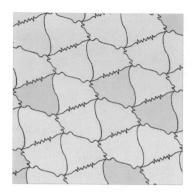

Heesch 04: The tiling

The symmetry group In the general case, the symmetry group is the *p2*.

The Heesch construction 05

Name TCTC.

The underlying network $(4, 4, 4, 4)$.

What can be specified? A free line, two C-lines.

Construction description It starts like *TTTT* with a free line from *A* to *B*, which is shifted to a line from *C* to *D*. Then *A* and *D* as well as *B* and *D* are each connected by a *C* line.

Number of copies of the fundamental domain in the translation scheme Two.

The fundamental domain and the translation scheme

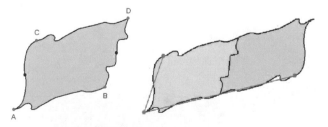

Heesch 05: Fundamental domain and translation scheme

The corresponding tiling

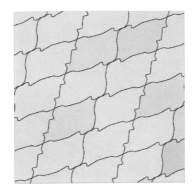

Heesch 05: The tiling

The symmetry group In the general case, the symmetry group is the *p2*.

The Heesch construction 06

Name TTTT.

The underlying network $(4, 4, 3, 3, 3)$.

What can be specified? A free line, three C-lines.

Construction description Draw a free line from *A* to *B* and move it to a line from *C* to D. Choose any other point *E* and then connect *A* and *C*, *B* and *E*, *E* and *D* each with a *C*-line.

Number of copies of the fundamental domain in the translation scheme Two.

The fundamental domain and the translation scheme

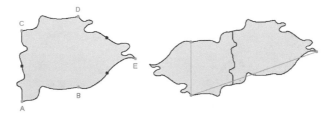

Heesch 06: Fundamental domain and translation scheme

The associated tiling

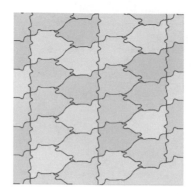

Heesch 06: The tiling

The symmetry group In the general case, the symmetry group is the *p2*.

The Heesch construction 07

Name TCCTCC.

The underlying network (3, 3, 3, 3, 3, 3).

What can be specified? A free line, four C-lines.

Construction description Draw a free line from *A* to *B* and move it to a line from *C* to *D*. Choose any two additional points *E* and *F* then connect *A* and *E*, *E* and *C*, *B* and F, *F* and *D* each with a *C* line.

Number of copies of the fundamental domain in the translation scheme One.

The fundamental domain and the translation scheme

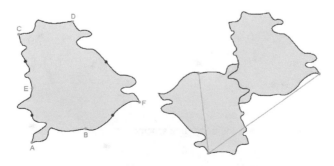

Heesch 07: Fundamental domain and translation scheme

The associated tiling

Heesch 07: The tiling

The symmetry group In the general case, the symmetry group is the *p2*.

The Heesch construction 08

Name $C_3 C_3 C_3 C_3$.

The underlying network $(6, 3, 6, 3)$.

What can be specified? Two free lines.

Construction description Draw a free line from A to B and rotate it by $120°$ around B; the endpoint of the rotated line is C. Reflect B at the line AC, which is point D. Then draw a second free line from A to D and rotate it (by D) by an angle of $120°$. This then connects D and C.

Number of copies of the fundamental domain in the translation scheme Three.

The fundamental domain and the translation scheme

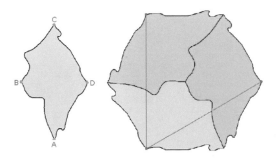

Heesch 08: Fundamental domain and translation scheme

The associated tiling

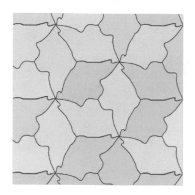

Heesch 08: The tiling

The symmetry group In the general case, the symmetry group is the *p*3.

The Heesch construction 09

Name $C_3 C_3 C_3 C_3 C_3 C_3$.

The underlying network $(3, 3, 3, 3, 3, 3)$.

What can be specified? Three free lines.

Construction description Rotate the free line from *A* to *B* by 120° around *A* to *C*. Choose any point *D* and draw a free line from *C* to *D*, which is then rotated by *D* by 120° to *E*. Determine *F* so that *ADF* forms an equilateral triangle. Then there is still a free line from *E* to *F* to be drawn, which is rotated by *F* by 120° to *B*.

Number of copies of the fundamental domain in the translation scheme Three.

The fundamental domain and the translation scheme

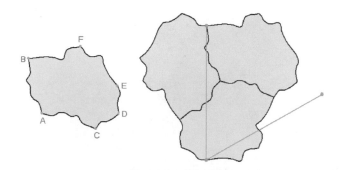

Heesch 09: Fundamental domain and translation scheme

The associated tiling

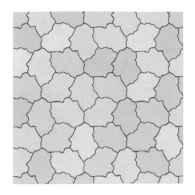

Heesch 09: The tiling

The symmetry group In the general case, the symmetry group is the *p*3.

The Heesch construction 10

Name CC$_3$*C*$_3$.

The underlying network (12, 12, 3).

What can be specified? A *C*-line and a free line.

Construction description It begins with a free line from *A* to *B*, which is rotated by *B* by 120° to *C*. Then connect *A* and *C* with a *C*-line.

Number of copies of the fundamental domain in the translation scheme Six.

The fundamental domain and the translation scheme

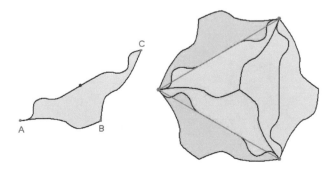

Heesch 10: Fundamental domain and translation scheme

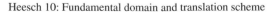

The corresponding tiling

Heesch 10: The tiling

The symmetry group In the general case, the symmetry group is the *p*6.

The Heesch construction 11

Name CC_6C_6.

The underlying network $(6, 6, 6)$.

What can be specified? A *C*-line and a free line.

Construction description The instruction is as in the previous case, but this time only to rotate by 60°.

Number of copies of the fundamental domain in the translation scheme Six.

The fundamental domain and the translation scheme

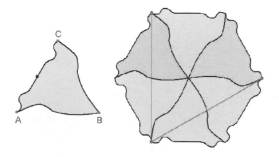

Heesch 11: Fundamental domain and translation scheme

The corresponding tiling

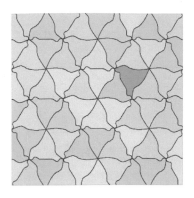

Heesch 11: The tiling

The symmetry group In the general case, the symmetry group is the *p*6.

The Heesch construction 12

Name $C_3C_3C_6C_6$.

The underlying network $(6, 3, 6, 3)$.

What can be specified? Two free lines.

Construction description Draw a free line from *A* to *D* and rotate it by 120° around *D* to *C*. Choose *B* so that the points *A*, *B* and *C* form an equilateral triangle. Draw a free line from *A* to *B* and rotate it by 60° around *B* to *C*.

Number of copies of the fundamental domain in the translation scheme Six.

The fundamental domain and the translation scheme

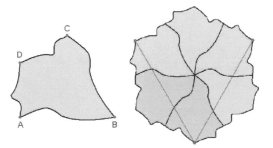

Heesch 12: Fundamental domain and translation scheme

The associated tiling

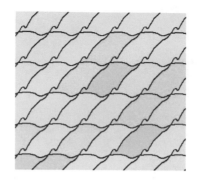

Heesch 12: The tiling

The symmetry group In the general case, the symmetry group is the *p*6.

The Heesch construction 13

Name $CC_3C_3C_6C_6$.

The underlying network $(6, 3, 3, 3, 3)$.

What can be specified? Two free lines and a *C*-line.

Construction description It begins with a free line from *A* to *B*, which is rotated by 120° to *A* to *C*. Now any point *D* is to be chosen, and *B* and *D* are to be connected by a free line. This free line is then rotated by 60° around *D* to *E*, and finally *E* and *C* are to be connected by a *C*-line.

Number of copies of the fundamental domain in the translation scheme Six.

The fundamental domain and the translation scheme

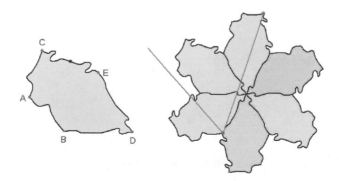

Heesch 13: Fundamental domain and translation scheme

The associated tiling

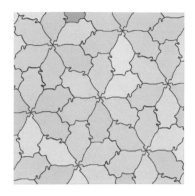

Heesch 13: The tiling

The symmetry group In the general case, the symmetry group is the *p6*.

The Heesch construction 14

Name CC_4C_4.

The underlying network $(8, 8, 4)$.

What can be specified? A free line and a C-line.

Construction description It is similar to the instructions 10 and 11, but this time the rotation angle is 90°.

Number of copies of the fundamental domain in the translation scheme Four.

The fundamental domain and the translation cell

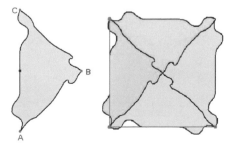

Heesch 14: Fundamental domain and translation cell

The corresponding tiling

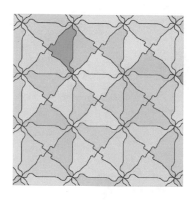

Heesch 14: The tiling

The symmetry group In the general case, the symmetry group is the *p*4.

The Heesch construction 15

Name $C_4 C_4 C_4 C_4$.

The underlying network $(4, 4, 4, 4)$.

What can be specified? A free line and a *C*-line.

Construction description The points *A*, *B*, *C* and *D* form the corners of a square, the lines are drawn as follows: From *A* to *B* is a free line, which is rotated by *B* to *D*, and the free line from *A* to *C* is rotated by *C* also to *D*.

Number of copies of the fundamental domain in the translation scheme Four.

The fundamental domain and the translation scheme

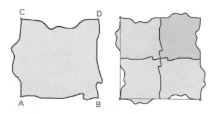

Heesch 15: Fundamental domain and translation scheme

The associated tiling

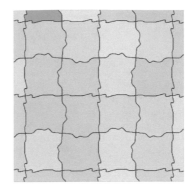

Heesch 15: The tiling

The symmetry group In the general case, the symmetry group is the *p*4.

The Heesch construction 16

Name $CC_4C_4C_4C_4$.

The underlying network $(4, 3, 4, 3, 3)$.

What can be specified? Two free lines and a *C* line.

Construction description The construction begins with a free line from *A* to *B*, which is rotated by 90° to *B* to *C*. Then another point *D* is chosen and connected to *A* with a free line. This line is rotated by 90° around *D*, the endpoint is the point *E*. Connect *E* and *C* by a *C* line.

Number of copies of the fundamental domain in the translation scheme Four.

The fundamental domain and the translation scheme

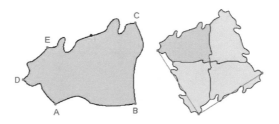

Heesch 16: Fundamental domain and translation scheme

The associated tiling

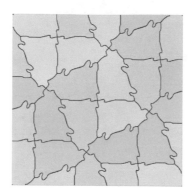

Heesch 16: The tiling

The symmetry group In the general case, the symmetry group is the *p*4.

The Heesch construction 17

Name $G_1G_1G_2G_2$.

The underlying network $(4, 4, 4, 4)$.

What can be specified? Two free lines.

Construction description Now glide reflections come into play. Draw a free line from A to B and carry out a glide reflection with it. The glide reflection axis G must have the same distance to A and B. After the glide reflection, A should be connected to the new point C.

Choose D so that the line AD is perpendicular to G and connect B and D with a free line, which then connects G and D by a glide reflection (on an axis parallel to C).

Number of copies of the fundamental domain in the translation grid Two.

The fundamental domain and the translation cell

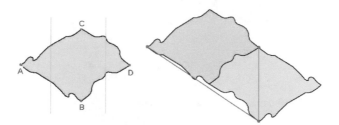

Heesch 17: Fundamental domain and translation cell

The corresponding tiling

Heesch 17: The tiling

The symmetry group In the general case, the symmetry group is the *pg*.

The Heesch construction 18

Name $TG_1G_1TG_2G_2$.

The underlying network $(3, 3, 3, 3, 3, 3)$.

What can be specified? Three free lines.

Construction description We start with a free line from B to D, which is shifted to a connecting line from C to E. Then choose a point A on the median of the line B C and connect A and B by a free line. By a glide reflection, a connection from A to C results.
On the right side this is repeated: Choose point F on the median of DE, connect with B by a free line and then close the figure by a glide reflection.

Number of copies of the fundamental domain in the translation scheme Two.

The fundamental domain and the translation cell

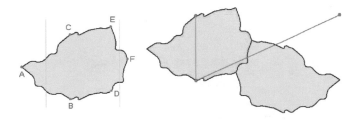

Heesch 18: Fundamental domain and translation cell

The corresponding tiling

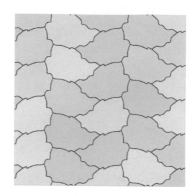

Heesch 18: The tiling

The symmetry group In the general case, the symmetry group is the *pg*.

The Heesch construction 19

Name TGTG.

The underlying network (4, 4, 4, 4).

What can be specified? Two free lines.

Construction description The beginning is as in the very first example: A free line from *A* to *B* is shifted to a connection from *C* to *D*. Now *A* and *C* are connected by a free line, which then becomes a connecting line from *B* to *D* by glide reflection. For this, the mirror axis must have the same distance to *A* and *D*.

Number of copies of the fundamental domain in the translation scheme Two.

The fundamental domain and the translation scheme

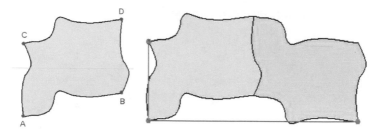

Heesch 19: Fundamental domain and translation scheme

The associated tiling

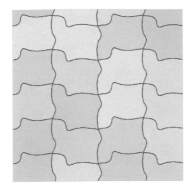

Heesch 19: The tiling

The symmetry group In the general case, the symmetry group is the *pg*.

The Heesch construction 20

Name $TG_1G_2TG_2G_1$.

The underlying network $(3, 3, 3, 3, 3, 3)$.

What can be specified? Three free lines.

Construction description The lines between A and B as well as C and D arise as in the previous example. Now an arbitrary point E is chosen and connected to A by a free line L. L is now to be shifted by a glide reflection so that L subsequently joins B; the (still) free endpoint is F. The glide reflection axis must be perpendicular to the line segment AC, it must have the same distance to E and B.
Now F is connected to D by a free line, by glide reflection a connection line from E to C arises. (The new glide reflection axis is parallel to the first, it has the same distance to C and F.)

Number of copies of the fundamental domain in the translation pattern Two.

The fundamental domain and the translation cell

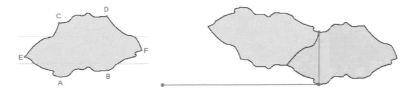

Heesch 20: Fundamental domain and translation cell

The corresponding tiling

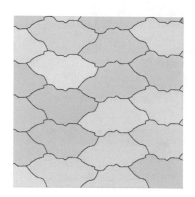

Heesch 20: The tiling

The symmetry group In the general case, the symmetry group is the *pg*.

The Heesch construction 21

Name CGG.

The underlying network (6, 6, 6).

What can be specified? Two free lines and a *C* line.

Construction description A and *B* are connected by a free line. Any line that has the same distance to *A* and *B* is used as a glide mirror axis, the free line is transformed on it so that a connecting line from *B* to *C* is created. *A* and *C* are still connected by a *C* line.

Number of copies of the fundamental domain in the translation scheme Four.

The fundamental domain and the translation scheme

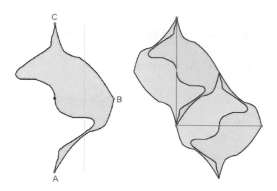

Heesch 21: Fundamental domain and translation scheme

The associated tiling

Heesch 21: The tiling

The symmetry group In the general case, the symmetry group is the *pgg*.

The Heesch construction 22

Name CCGG.

The underlying network $(4, 4, 4, 4)$.

What can be specified? A free line and two *C*-lines.

Construction description The first three points *A*, *B* and *C* arise as in the previous example. This time, another arbitrary point *D* is added, we connect *A* and *D* and *D* and *C* each with a *C*-line.

Number of copies of the fundamental domain in the translation scheme Four.

The fundamental domain and the translation cell

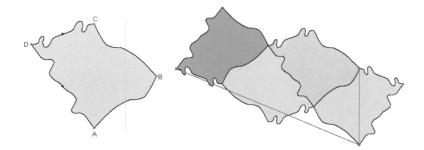

Heesch 22: Fundamental domain and translation cell

The corresponding tiling

Heesch 22: The tiling

The symmetry group In the general case, the symmetry group is the *pgg*.

The Heesch construction 23

Name TCTGG.

The underlying network $(4, 4, 3, 3, 3)$.

What can be specified? Two free lines and a *C*-line.

Construction description As already several times before, a free line from *A* to *B* is shifted to a connecting line from *C* to *D*. On the middle perpendicular of the section *BD*, *E* is selected and initially connected to a free line with *B* and then by glide reflection to a connection from *E* to *D*. It only remains to connect *A* and *C* with a *C* -line.

Number of copies of the fundamental domain in the translation scheme Four.

The fundamental domain and the translation scheme

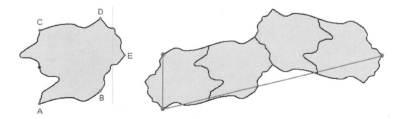

Heesch 23: Fundamental domain and translation scheme

The associated tiling

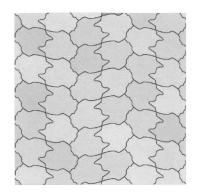

Heesch 23: The tiling

The symmetry group In the general case, the symmetry group is the *pgg*.

The Heesch construction 24

Name TCCTGG.

The underlying network $(3, 3, 3, 3, 3, 3)$.

What can be specified? Two free lines and two *C*-lines.

Construction description The first steps—up to the connection of *A* with *C*—are as in the previous example. Now another point *F* is chosen, and *A* and *F* as well as *F* and *C* are each connected by a *C*-line.

Number of copies of the fundamental domain in the translation scheme Four.

The fundamental domain and the translation scheme

Heesch 24: Fundamental domain and translation scheme

The corresponding tiling

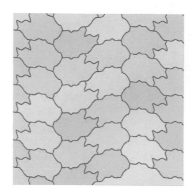

Heesch 24: The tiling

The symmetry group In the general case, the symmetry group is the *pgg*.

The Heesch construction 25

Name CGCG.

The underlying network $(4, 4, 4, 4)$.

What can be specified? A free line and two *C*-lines.

Construction description The free line from *A* to *B* is reflected by glide reflection to a connection from *C* to *D*. The glide reflection axis has the same distance to *A* and *D*. Connect *A* and *C* as well as *B* and *D* by a *C*-line.

Number of copies of the fundamental domain in the translation scheme Four.

The fundamental domain and the translation scheme

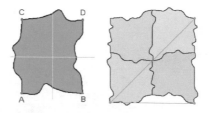

Heesch 25: Fundamental domain and translation scheme

The associated tiling

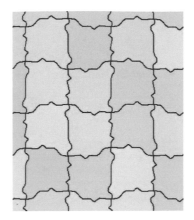

Heesch 25: The tiling

The symmetry group In the general case, the symmetry group is the *pgg*.

The Heesch construction 26

Name $G_1 G_2 G_1 G_2$

The underlying network $(4, 4, 4, 4)$.

What can be specified? Two free lines.

Construction description The points A, B, C, D are the corners of a rectangle. A free line from A to B connects by glide reflection the points C and D, and another free line from A to C provides by glide reflection a connection from B to D. The two glide reflection axes are orthogonal and parallel to the sides of the rectangle.

Number of copies of the fundamental domain in the translation scheme Four.

The fundamental domain and the translation scheme

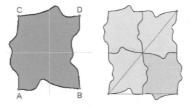

Heesch 26: Fundamental domain and translation scheme

The associated tiling

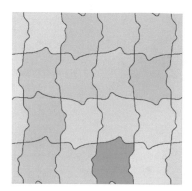

Heesch 26: The tiling

The symmetry group In the general case, the symmetry group is the *pgg*.

The Heesch construction 27

Name $CG_1G_2G_1G_2$.

The underlying network $(4, 3, 4, 3, 3)$.

What can be specified? Two free lines and an *C*-line.

Construction description The points A, B, C, D and the connecting lines from A to B and C to D arise as in the penultimate example 25. Now B is connected to D by a free line L.
Now it gets a little more complicated: You have to find a glide reflection that transforms L so that the line afterwards joins at A (the endpoint should be called E). For this, the glide reflection axis must be perpendicular to the one from the first step. Also connect E and C by a C line.

Number of copies of the fundamental domain in the translation scheme Four.

The fundamental domain and the translation scheme

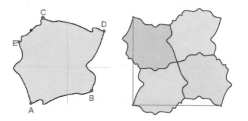

Heesch 27: Fundamental domain and translation scheme

The corresponding tiling

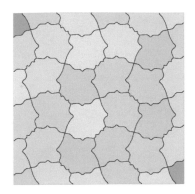

Heesch 27: The tiling

The symmetry group In the general case, the symmetry group is the *pgg*.

The Heesch construction 28

Name $CG_1 CG_2 G_1 G_2$.

The underlying network $(3, 3, 3, 3, 3, 3)$.

What can be specified? Two free lines and two *C*-lines.

Construction description It starts again like in Example 25: The connection of *C* to *D* is created by glide reflection (axis: *G*) from a free line from *A* to *B*. Now choose a free line from *B* to *D* and transform it by glide reflection to a line from *E* to *F*. The glide reflection axis is perpendicular to *G*. Finally, connect *A* and *E* as well as *F* and *C* by a *C* line.

Number of copies of the fundamental domain in the translation scheme Four.

The fundamental domain and the translation cell

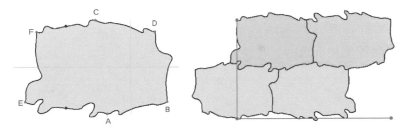

Heesch 28: Fundamental domain and translation cell

The associated tiling

Heesch 28: The tiling

The symmetry group In the general case, the symmetry group is the *pgg*.

References for Part I

The following are some selected *references for Part I (Escher)*: literature that played a special role in the preparation of this book.

Behrends, Ehrhard: Escher über die Schulter gesehen. In "Pi and Co - Kaleidoscope of Mathematics", Springer, 2016. Edited by E. Behrends, P. Gritzmann, G. Ziegler, 2nd edition, about 420 pages.
This article already described the Heesch constructions in detail.

Bigalke, Hans-Günther: Heinrich Heesch: Kristallgeometrie, Parkettierungen, Vierfarbenforschung. Birkhäuser, 1988.
A recommended biography.

Heesch, Heinrich and Kienzle, Otto: Flächenschluss. Springer, 1963.
A classic: Here Heesch and Kienzle describe the Heesch constructions for engineers.

Martin, George E.: The Foundations of Geometry and the Non-Euclidean Plane. Springer, 1975.
For those interested in non-Euclidean geometries.

Martin, George E.: Transformation Geometry (an Introduction to Symmetry). Springer, 1982.
A standard work on plane geometry. You will also find a quite brief derivation of the Frieze groups and the plane crystal groups.

Schattschneider, Doris: M. C. Escher – Visions of Symmetry. Freeman and Company, 1990.
Highly recommended: Everything about Escher and his work, written from the perspective of a mathematician for interested laymen.

Ernst, Bruno: Abenteuer mit unmöglichen Figuren. Taco Verlag, 1987.
Here, another aspect of Escher's work is dealt with: the impossible figures.

    It should also be noted that the **Internet** is a rich source of information on the topic of "symmetry". I would particularly like to recommend the websites

    https://itunes.apple.com/de/app/iornament/id534529876?mt=8;
    https://itunes.apple.com/de/app/iornament-crafter-3d-basteln-muster-und-mehr/id1183533412?mt=8
    of my colleague Jürgen Richter-Gebert from Munich.

Part II
Möbius Transformations

Möbius Transformations

5

In the second part of the book we study Möbius transformations[1]. They are named after *August Ferdinand Möbius (1790–1868)*. He was director of the Leipzig Observatory and professor of mathematics and astronomy at the local university. Even the *Möbius band*, a non-orientable surface in $\mathbb{R}^3$, bears his name. It was described independently by him and Johann Benedict Listing in 1858.

In this and the following chapter – after a brief reminder of some facts from the theory of functions – Möbius transformations are defined and examined. Möbius transformations are special maps of the complex plane into itself, they correspond to the movements of the plane in the previous chapter[2]. We first study some basic properties of such transformations. The aim of this chapter is a complete characterization: They can be divided into *four classes*, the *elliptic, hyperbolic, parabolic and loxodromic* transformations.

The more important and interesting results will then be in the next chapter. In particular, we will investigate connections to hyperbolic geometry and describe analogies to "classical" tilings of the plane à la Escher in this geometry.

5.1   Complex Numbers: Some Reminders

For the following considerations, only relatively basic properties of complex numbers and functions defined between sets of complex numbers play a role. Many of the definitions and results are already known from analysis, some theorems from complex analysis are taken over as building blocks.

---

[1] In the classical literature they are also called "broken linear transformations", today one almost exclusively speaks of Möbius transformations.

[2] However, this is only a very vague comparison. Both movements of the plane and Möbius transformations are characterized by certain invariance properties, and later we will again be interested in groups of such maps.

© The Author(s), under exclusive license to Springer Fachmedien Wiesbaden GmbH, part of Springer Nature 2022
E. Behrends, *Tilings of the Plane*, Mathematics Study Resources 2,
https://doi.org/10.1007/978-3-658-38810-2_5

The set $\mathbb{C}$ of complex numbers
One can imagine $\mathbb{C}$ as the Cartesian plane $\mathbb{R} \times \mathbb{R}$, and define addition and multiplication on it by the following formulas:

$$(x, y) + (x', y') := (x + x', y + y'), \quad (x, y) \cdot (x', y') := (xx' - yy', xy' + yx').$$

If one denotes $(0, 1)$ as i and (x, y) always as $x + iy$, then one only needs to remember: The arithmetic is like with real numbers, and if i^2 appears anywhere, one can replace it with -1.

Then the field axioms apply in $\mathbb{C}$, and it can be shown that $\mathbb{C}$ cannot be ordered. $|x + iy|$ is defined as $\sqrt{x^2 + y^2}$, and then $d(z, w) := |z - w|$ gives rise to a metric which leads to a definition of convergence. $\mathbb{C}$ becomes a complete metric space as a result.

The completion of $\mathbb{C}$ $\hat{\mathbb{C}}$ is defined as $\mathbb{C} \cup \{\infty\}$. Neighbourhoods of ∞ are all subsets $U \subset \hat{\mathbb{C}}$ that contain ∞ and the complement of a circle with center 0: There should be an $R > 0$ so that $\{z \mid |z| > R\} \subset U$. A sequence (z_n) of complex numbers is then obviously convergent to ∞ if, for every $R > 0$, there is an n_0 so that $|z_n| > R$ for $n \geq n_0$.

In other words, all points "far out" are close to ∞, and therefore it has proven useful to imagine $\hat{\mathbb{C}}$ as a sphere: $\mathbb{C}$ is "bent up" and above completed by ∞ to a sphere.

Differentiable functions If O is an open subset of $\mathbb{C}$, $z_0 \in O$ and $f : O \to \mathbb{C}$ is a function, then f is differentiable at z_0 with derivative $f'(z_0)$ if always

$$\frac{f(z_n) - f(z_0)}{z_n - z_0} \to f'(z_0)$$

for sequences z_n with $z_n \to z_0$ holds, where $z_n \neq z_0$ for all n. Differentiability is thus defined exactly as in $\mathbb{R}$. A function that is differentiable at all $z_0 \in O$ is called *analytic* or also *holomorphic*. Surprisingly, holomorphic functions behave much "more regular" than differentiable functions that are defined on subsets of $\mathbb{R}$. Here are some examples of this phenomenon:

- If holomorphic functions agree on O on a sequence (z_n) that has an accumulation point in O, then they are identical if O is connected.
- If $f : \mathbb{C} \to \mathbb{C}$ is holomorphic and bounded, then f is constant (Liouville's theorem).
- Holomorphic functions are differentiable arbitrarily often and can be expanded locally into a power series at each point.

Using obvious calculation rules for ∞, one can expand the algebraic and topological definitions and "differentiability" on $\hat{\mathbb{C}}$. So is $a/\infty := 0$, $\infty + a := a + \infty := \infty$ for all $a \in \mathbb{C}$ etc. And is $O \subset \hat{\mathbb{C}}$ with $\infty \in O$ open and $f : O \to \mathbb{C}$ a function, so f is called continuous/differentiable at ∞, if $z \mapsto f(1/z)$ is continuous/differentiable at 0. So is for example $1/z$ everywhere on $\hat{\mathbb{C}} \setminus \{0\}$ differentiable.

5.2 Möbius Transformations: Definitions and First Results

This chapter is based on the following definitions. For $a, b, c, d \in \mathbb{C}$, a mapping $M_{a,b,c,d}$ is defined by

$$z \mapsto \frac{az + b}{cz + d}$$

To ensure that the denominator is not always 0, c and d should not be zero at the same time. Also, the case $ad = bc$ would be uninteresting, because then the mapping would be constant with the value a/c or b/d [3]. In short: We will require that $ad - bc \neq 0$, , then both requirements are met.

It is also clear that one obtains the same mapping if one replaces a, b, c, d by ka, kb, kc, kd for an arbitrary $k \neq 0$. If one chooses k so that $k^2(ac - bd) = 1$ holds, this is no restriction. We define:

Definition 5.2.1
(i) Let $a, b, c, d \in \mathbb{C}$ with $ac - bd \neq 0$. Under the associated *Möbiustransformation* one understands the mapping

$$M_{a,b,c,d} : z \mapsto \frac{az + b}{cz + d}.$$

(ii) The Möbiustransformation is called *normalized*, if $ad - bc = 1$ holds.

It makes sense to choose the domain and the range as $\hat{\mathbb{C}}$:

- If $c \neq 0$, then $M_{a,b,c,d}(-d/c) := \infty$ and $M_{a,b,c,d}(\infty) := a/c$.
- In the case of $c = 0$ (then $a \neq 0 \neq d$), we define $M_{a,b,c,d}(\infty) := \infty$.

In the sequel we will investigate many *examples* in more detail, first think of simple functions like $z, 2z + i, iz, 1/z, \ldots$

Lemma 5.2.2
Möbius transformations can be inverted: The mapping

$$M_{d,-b,-c,a} : z \mapsto \frac{dz - b}{-cz + a}$$

is inverse to $M_{a,b,c,d}$.

---

[3] For example, let $d \neq 0$. From $ad = bc$ it follows that $d(az + b) = b(cz + d)$ for all z, thus $(az + b)/(cz + d) = b/d$ for all $z \neq -d/c$. And if $c \neq 0$, then from $c(az + b) = a(cz + b)$ it follows that $(az + b)/(cz + d) = a/c$ for these z.

Proof Let $M := M_{a,b,c,d}$ and $\hat{M} := M_{d,-b,-c,a}$. It has to be shown that always $M \circ \hat{M}(z) = z$ and $\hat{M} \circ M(z) = z$ applies. We calculate as follows:

$$
\begin{aligned}
M \circ \hat{M}(z) &= \frac{a\frac{dz-b}{-cz+a} + b}{c\frac{dz-b}{-cz+a} + d} \\
&= \frac{a(dz-b) + b(-cz+a)}{c(dz-b) + d(-cz+a)} \\
&= \frac{z(ad-bc)}{ad-bc} \\
&= z.
\end{aligned}
$$

There are still some special cases to be considered. Let, for example, $c \neq 0$. Then $\hat{M}(a/c) = \infty$ and $M(\infty) = a/c$, so it is really $M \circ \hat{M}(a/c) = a/c$. And $\hat{M}(\infty) = -d/c$, so it also applies $M \circ \hat{M}(\infty) = \infty$. And in the case of $c = 0$, M and $\hat{M}$ map the value ∞ on itself.

Quite analogously one shows that $\hat{M} \circ M(z) = z$ for all z.   □

Remarks

1. It follows that with $M_{a,b,c,d}$ also the inverse mapping is normalized.
2. Also $M_{-d,b,c,-a}$ is inverse to $M_{a,b,c,d}$. Sometimes it is important to choose the appropriate inverse mapping.

The set of Möbius transformations forms a group:

Proposition 5.2.3
(i) With $M := M_{a,b,c,d}$ and $\hat{M} := M_{a',b',c',d'}$ also $\hat{M} \circ M$ is a Möbius transformation. If both are normalized, then so is the product.
(ii) The set of all Möbius transformations forms a subgroup of the group of bijective maps from $\hat{\mathbb{C}}$ to $\hat{\mathbb{C}}$ (with the mapping composition as group multiplication).

Proof (i) For arbitrary z one has

$$
\begin{aligned}
M \circ \hat{M}(z) &= \frac{a\frac{a'z+b'}{c'z+d'} + b}{c\frac{a'z+b'}{c'z+d'} + d} \\
&= \frac{a(a'z+b') + b(c'z+d')}{c(a'z+b') + d(c'z+d')} \\
&= \frac{z(aa'+bc') + (ab'+bd')}{z(ca'+dc') + (cb'+dd')}.
\end{aligned}
$$

Then it remains to be noted that

$$(aa' + bc')(cb' + dd') - (ab' + bd')(ca' + dc') = (ad - bc)(a'd' - b'c')$$

holds. If therefore $ad - bc, a'd' - b'c'$ are not equal to zero (or even equal to one), then $(aa' + bc')(cb' + dd') - (ab' + bd')(ca' + dc')$ is not equal to zero (or even equal to one).

(ii) This follows from (i) and Lemma 5.2.2.   □

As a side result, we can make an interesting connection to matrices. Denoting as usual by $SL(2, \mathbb{C})$ the complex 2×2 -matrices with determinant 1 and considering the mapping $\left(\begin{smallmatrix} a & b \\ c & d \end{smallmatrix} \right) \mapsto M_{a,b,c,d}$, this is a group homomorphism from $SL(2, \mathbb{C})$ to the normalized Möbius transformations. Therefore, due to elementary theorems about matrices, it is not surprising afterwards that inverses and products of normalized transformations are again normalized. The kernel of this mapping consists, as is easily seen, of the matrices $\left(\begin{smallmatrix} a & 0 \\ 0 & 1/a \end{smallmatrix} \right)$ with $a \neq 0$.

The proposition implies that all $M_{a,b,c,d}$ are one-to-one if one restricts their domain to $\mathbb{C}$ (with – possibly – the exception of $-d/c$), and they are obviously also holomorphic there. By this property, they are already characterized:

Proposition 5.2.4
Let $f : \hat{\mathbb{C}} \to \hat{\mathbb{C}}$ be holomorphic and one-to-one[4]. Then f is a Möbius transformation.

Proof First, let $f(0) = 0$ and $f(\infty) = \infty$. f is therefore an entire function, i.e. f can be written as a power series $f(z) = a_1 z + a_2 z^2 + \cdots$ which converges everywhere. If f were *not* a polynomial, then ∞ would be an essential singularity. According to the Casorati-Weierstrass theorem, this would then map every neighbourhood of ∞ onto a dense subset of $\mathbb{C}$, f would therefore not be continuous at ∞. f is therefore a polynomial. But the only injective polynomial, due to the fundamental theorem of algebra, has the form $a_1 z$ with $a_1 \neq 0$.

In the general case, we compose f with Möbius transformations M_1, M_2 so that for $M_1 \circ M_2 \circ f$ the conditions of the first part of the proof are satisfied. This is then, as already shown, a Möbius transformation M, and we obtain $f = M_2^{-1} \circ M_1^{-1} \circ M$. This is, based on the already proven results, again a Möbius transformation.

To define M_1, M_2, we set $z_0 := f(\infty)$ and $w_0 := f(0)$, and we consider the following cases:

Case 1: $z_0 = \infty, w_0 \in \mathbb{C}$. Set $M_1 = \text{Id}$ and $M_2(z) = z - w_0$.

---

[4] Just as a reminder: Holomorphicity for ∞ should mean that $z \mapsto f(1/z)$ is holomorphic at 0.

Case 2: $z_0 \in \mathbb{C}, w_0 = \infty$. Set $M_1 = \mathrm{Id}$ and $M_2(z) = 1/(z - z_0)$.

Case 3: $z_0, w_0 \in \mathbb{C}$. Set $M_1(z) = z - 1/(w_0 - z_0)$ and $M_2(z) = 1/(z - z_0)$.

It should be noted that because of the injectivity of f, case $z_0 = w_0$ cannot occur.   □

If we write $M_{a,b,c,d}$ in case $c \neq 0$ as

$$
\frac{az + b}{cz + d} = \frac{(a/c)(cz + d) + (b - ad/c)}{cz + d}
$$
$$
= \frac{a}{c} + \frac{b - ad/c}{cz + d},
$$

we see that $M_{a,b,c,d}$ is composed of simple building blocks:

Lemma 5.2.5

(i) Let $c \neq 0$. Then we can write $M_{a,b,c,d}$ as

$$
M_{a,b,c,d} = M_5 \circ M_4 \circ M_3 \circ M_2 \circ M_1;
$$

where M_1, M_2, M_3, M_4, M_5 are the following Möbius transformations:

$$
M_1(z) := cz; \quad M_2(z) := z + d;
$$
$$
M_3(z) := \frac{1}{z}; \quad M_4(z) := \left(b - \frac{ad}{c}\right)z; \quad M_5(z) := \frac{a}{c} + z.
$$

(ii) In the case $c = 0$, $M_{a,b,c,d} = M_2 \circ M_1$, where

$$
M_1(z) := \frac{a}{d}z; \quad M_2(z) := z + \frac{b}{d}.
$$

This means that, in order to understand the general properties of Möbius transformations, one only needs to study the maps

$$
z \mapsto \alpha z, \; z \mapsto z + \beta, \; z \mapsto \frac{1}{z}
$$

(with $\alpha, \beta \in \mathbb{C}, \alpha \neq 0$). For example, it immediately follows once again that all $M_{a,b,c,d}$ are bijective maps on $\hat{\mathbb{C}}$.

5.3   Möbius Transformations and Circles

Circles play a special role when studying Möbius transformations. It is also reasonable to consider lines as circles: In $\hat{\mathbb{C}}$ one can consider lines as circles that pass through ∞. We will use the following terminology:

Definition 5.3.1

(i) Let $z_0 \in \mathbb{C}$ and $r > 0$. Then the *circle around z_0 with the radius r* is defined by

$$K_{z_0,r} := \{z \in \mathbb{C} \mid |z - z_0| = r\}.$$

For each circle z_0 and r are uniquely determined.

(ii) Let G be a line in $\mathbb{C}$ that does not pass through zero. There is then a uniquely determined $z_0 \in \mathbb{C}, z_0 \neq 0$ so that G can be written as

$$G_{z_0,iz_0} := \{z_0 + tiz_0 \mid t \in \mathbb{R}\}.$$

Lines that go through zero are written as

$$G_{0,z_1} := \mathbb{R}\,z_1,$$

where $z_1 \neq 0$. In all cases, therefore, $w_0, w_0 + w_1$ are different points of the line G_{w_0,w_1}, which uniquely determine it.

In general, we understand (if $z_1 \neq 0$) under G_{z_0,z_1} the line $z_0 + \mathbb{R}\,z_1$.

We claim that Möbius transformations map circles and lines to circles and lines. Because of Lemma 5.2.5 we only have to prove this for the *simple* transformations listed there:

Lemma 5.3.2

(i) Let $M(z) := \alpha z$ with $\alpha \neq 0$. Then

$$M(K_{z_0,r}) = K_{\alpha z_0,|\alpha|r}, \ M(G_{z_0,z_1}) = G_{\alpha z_0,\alpha z_1}.$$

(ii) Let $M(z) := z + \beta$ with $\beta \in \mathbb{C}$. Then

$$M(K_{z_0,r}) = K_{z_0+\beta,r}, \ M(G_{z_0,z_1}) = G_{z_0+\beta,z_1}.$$

(iii) Let $M(z) = 1/z$.

(iii.1) If $K = K_{z_0,r}$ and 0 *not* lies on the boundary of K, then $M(K)$ is a circle again: The center is $w_0 := \overline{z_0}/(z_0\overline{z_0} - r^2)$ and the radius is

$$r' := |r/(z_0\overline{z_0} - r^2)|.$$

(iii.2) If $K = K_{z_0,r}$ is a circle and 0 lies on the boundary of K (thus $|z_0| = r$), then $M(K)$ is the line G_{w_0,w_1} with $w_0 := 1/(2z_0)$, $w_1 := i\overline{z_0}$.

(iii.3) Let $G = G_{z_0,z_1}$ be a line that does *not* go through 0. Then $M(G)$ is a circle. More precisely: If one writes G in the form $\{w_0 + tiw_0 \mid t \in \mathbb{R}\}$ with a uniquely determined $w_0 \neq 0$, then the center of $M(G)$ is equal to $1/(2w_0)$ and the radius is equal to $1/|2w_0|$.

(iii.4) If $G = G_{0,z_1}$ is a line going through zero, then $M(G)$ is the line $G_{0,1/z_1}$.

Proof (i), (ii) and (iii.4) are clear.

(iii.1) Let $z \in K_{z_0,r}$, it follows that $(z - z_0)\overline{(z - z_0)} = r^2$. It has to be shown that

$$\left(\frac{1}{z} - \frac{z_0}{z_0 \overline{z_0} - r^2} \right) \left(\frac{1}{\overline{z}} - \frac{\overline{z_0}}{z_0 \overline{z_0} - r^2} \right) = \left(\frac{r}{z_0 \overline{z_0} - r^2} \right)^2.$$

We transform this equation equivalently by multiplying it with $z \overline{z}(z_0 \overline{z_0} - r^2)^2$ (according to the assumption this number is different from zero). Then we get

$$\left((z_0 \overline{z_0} - r^2) - \overline{z_0} z \right) \left((z_0 \overline{z_0} - r^2) - \overline{z} z_0 \right) = r^2 z \overline{z}.$$

An easy calculation gives rise to

$$(z_0 \overline{z_0} - r^2)^2 + z_0 \overline{z_0} z \overline{z} - (z_0 \overline{z_0} - r^2)2 \operatorname{Re}(z_0 \overline{z}) = r^2 z \overline{z}.$$

$(z_0 \overline{z_0} - r^2)$ can be cancelled, it remains

$$(z_0 \overline{z_0} - r^2) + z \overline{z} - 2 \operatorname{Re}(z_0 \overline{z}) = 0,$$

and this states $(z - z_0)\overline{(z - z_0)} = r^2$ in disguise. With that $M(K) \subset K_{w_0,r'}$ is proven.

And conversely, the same also holds: If $w \in K_{w_0,r'}$, then $w \in M(K)$. For that one shows $1/w \in K$, which can be easily verified with an analogous calculation.

(iii.2) First let z_0 be real. We choose z with $|z - z_0| = r = |z_0|$ and examine $1/z$. According to the assumption $(z - z_0)\overline{(z - z_0)} = z_0 \overline{z_0}$, so $z \overline{z} = 2 \operatorname{Re} z z_0 = 2 z_0 \operatorname{Re} z$. Next, we calculate the real part of $1/z$:

$$\begin{aligned}
\operatorname{Re} \frac{1}{z} &= \frac{1}{2}\left(\frac{1}{z} + \frac{1}{\overline{z}} \right) \\
&= \frac{z + \overline{z}}{2 z \overline{z}} \\
&= \frac{2 \operatorname{Re} z}{4 z_0 \operatorname{Re} z} \\
&= \frac{1}{2 z_0}.
\end{aligned}$$

(Note that z_0 is real.) Consequently, there is a $t \in \mathbb{R}$ with $1/z = 1/(2z_0) + it z_0$. Conversely, for all $z = 1/(2z_0) + it z_0$, it holds that $|1/z - z_0| = |z_0|$, because, as can be easily verified by multiplication with $\left(1/(2z_0) + it z_0 \right)\left(1/(2z_0) - it z_0 \right)$,

$$\left(\frac{1}{1/(2z_0 + it z_0) - z_0} \right) \left(\frac{1}{1/(2z_0 - it z_0) - z_0} \right) = z_0^2.$$

If z_0 is not real, we use part (i) of the lemma. For a suitable α, the center of αK is real. Consequently, the image of αK under $1/z$ is a line, and this image is the $1/\alpha$ -fold of the image of K.

(iii.3) This can be traced back to (iii.2). We write G in the form G_{w_0,w_1} with $w_0 := 1/(2z_0), w_1 := i\overline{z_0}$, i.e.

$$G = \left\{ \frac{1}{2z_0} + it\overline{z_0} \;\middle|\; t \in \mathbb{R} \right\}.$$

In (iii.2) we have shown that $G = M(K_{z_0,|z_0|})$ holds, and therefore – since M is inverse to itself – $M(G) = K_{z_0,|z_0|}$ follows as claimed. □

We now want to illustrate these results. In each case, you can see the unit circle (gray) and some lines and circles in different positions (blue). The respective image under the image $z \mapsto 1/z$ is drawn in red.

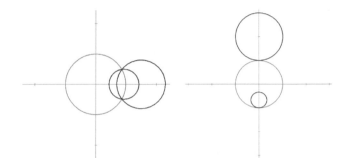

A circle with center in $\mathbb{R}$ and a circle with center on the imaginary axis

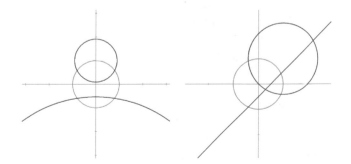

A circle that almost goes through zero and a circle through zero

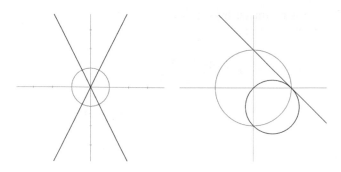

A line through zero and one that does not go through zero

At the end of this section, we will treat some *supplements on the topic of "circles"*.

Circles that intersect Assume that you want to map several circles under a Möbius transformation M. Let's say K_1, K_2 circles that intersect at an angle α in the point z_0: The cases $\alpha = 0$ and $\alpha = \pi/2$ are particularly important here. Then the (generalized) circles $M(K_1)$ and $M(K_2)$ will also intersect in $M(z_0)$ at the angle α. The justification: If a mapping f is holomorphic at z_0 with derivative $f'(z_0)$, different from zero, then f can be locally replaced at z_0 by $z \mapsto f(z_0) + f'(z_0)(z - z_0)$, and multiplication by $f'(z_0)$ corresponds to a rotation. This shows in particular:

> **Proposition 5.3.3**
> If K_1, K_2 are circles or lines or other curves that intersect orthogonally (resp. more generally at an angle ϕ), then $M(K_1)$ and $M(K_2)$ also intersect orthogonally (resp. at the same angle ϕ) for every Möbius transformation M.

Proof The justification was given essentially above. However, one still has to investigate the case when M is singular at the intersection point z_0, that is, if $cz_0 + d = 0$ holds. Then the intersection point is mapped to ∞. It is agreed: The intersection angle of two curves K_1, K_2 under a mapping $f : \hat{\mathbb{C}} \to \hat{\mathbb{C}}$ at ∞ is the intersection angle of $\hat{M}(K_1)$ and $\hat{M}(K_2)$ at zero under $z \mapsto f(1/z)$, where $\hat{M}$ denotes the Möbius transformation $z \mapsto 1/z$. The statement is easy to supplement for this case. □

How can you calculate the image circle quickly? If one wants to apply the above results to computer simulations, the question arises how to determine quickly and reliably $M(K)$ from K or $M(G)$ from G for circles K and lines G. I myself have had good experiences with the following two possibilities:

- A line or a circle is uniquely determined by three different points lying on it. So write two programs: The first generates three different points on the given line or the given circle. And the second determines the circle K or the line G with $z_1, z_2, z_3 \in K$ or $\in G$ for three different points z_1, z_2, z_3. Then proceed as follows:

If K or G are given, find suitable z_1, z_2, z_3 with program 1 that lie on it. Calculate the values $M(z_1), M(z_2), M(z_3)$ and apply program 2 to them.

- M can be written as a combination of very simple transformations (translation, rotation) and $z \mapsto 1/z$ (cf. Lemma 5.2.5). The most complicated is $1/z$, for which we have proved concrete formulas to describe the transformation of circles and lines. And for the "simple" transformations, it is obvious how the transformation formulas for circles and lines look (they also map circles into circles and lines into lines).

Interesting circle patterns If you form (generalized) circles that touch or intersect at an angle α under a Möbius transformation, the (generalized) image circles will also touch or intersect at an angle α, as we have shown. This can be used to create attractive images.

Here's the idea:

Step 1: Imagine a pattern of circles and lines that can be easily realized, for example by lines and circles with the same radius in the plane. Here are some examples where the circles were also drawn in different colors:

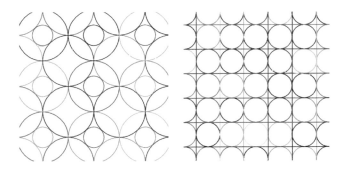

Two "simple circle patterns" in the plane

Step 2: Transform all circles and lines under a freely chosen Möbius transformation. The following images were created:

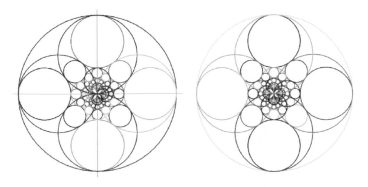

Circles, transformed with $1/z$

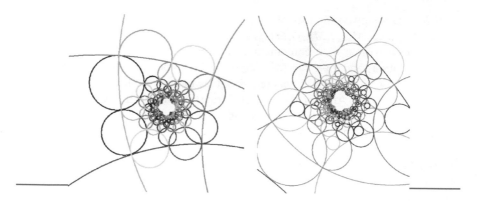

Circles, transformed with a more complicated Möbius transformation

5.4   Fixed Points of Möbius Transformations

Fixed points play an important role in the analysis of Möbius transformations. A $z \in \hat{\mathbb{C}}$ is called a *fixed point* of a Möbius transformation M if $M(z) = z$ holds. For example, the following holds true

- Every z is a fixed point of $z \mapsto z$ (the identical mapping).
- 0 and ∞ are fixed points of $z \mapsto (2 + i)z$.
- ∞ is the only fixed point of $z \mapsto z + 1$.
- 1 and -1 are the fixed points of $z \mapsto 1/z$.

This observation is a special case of the following theorem:

Proposition 5.4.1
Let $M = M_{a,b,c,d}$ be a normalized Möbius transformation (that is, $ad - bc = 1$ holds). The following cases are possible:

(i) M is the identity. Then all z fixed points.
(ii)   One has $c \neq 0$ and $a + d \in \{-2, 2\}$. Then there is exactly one fixed point, namely $(a - d)/2c$.
(iii) One has $c \neq 0$ and $a + d \notin \{-2, 2\}$. Then there are exactly two fixed points, namely

$$\frac{(a - d) \pm \sqrt{(a + d)^2 - 4}}{2c}.$$

(iv)   One has $c = 0$, so that $M(z) = (a/d)z + b/d$. If $b = 0$ and M is not the identity, then $a/d \neq 1$ must hold. (Necessarily, also $a/d \neq 0$.) There are two fixed points, namely 0 and ∞.

(v)   Suppose that $c = 0$, so $M(z) = (a/d)z + b/d$ again. If $b \neq 0$, two cases
are possible:
- It can be $a/d = 1$. Then ∞ is the only fixed point.
- If $a/d \neq 1$, there are the fixed points ∞ and $b/(b - a)$.

Proof Everything follows immediately from the explicit solution formula for
quadratic equations.   □

We still formulate the

Corollary 5.4.2
If $M(z) = z$ holds for three pairwise different z, then M is the identity.

It has an important consequence: If both z_1, z_2, z_3 and w_1, w_2, w_3 are pairs of different points in $\hat{\mathbb{C}}$, then one can find exactly one Möbius transformation that maps
these points into each other:

Proposition 5.4.3
There is exactly one M with $M(z_i) = w_i$ $((i = 1, 2, 3)$.

Proof The   uniqueness   is   clear: If $M, \hat{M}$ are   transformations   with
$M(z_i) = \hat{M}(z_i) = w_i$, then $M^{-1} \circ \hat{M}$ has the three fixed points z_i; due to the previous corollary, $M^{-1} \circ \hat{M}$ is the identity, it follows that $M = \hat{M}$.

For the existence proof, we first consider the case $w_1, w_2, w_3 = 0, 1, \infty$. First,
let all z_i be in $\mathbb{C}$. The Möbius transformation

$$M_{[z_1, z_2, z_3]}(z) := \frac{(z - z_1)(z_2 - z_3)}{(z - z_3)(z_2 - z_1)}$$

obviously has the desired properties. If a z_i, for example z_3, is equal to ∞, then

$$M_{[z_1, z_2, \infty]}(z) := \frac{z - z_1}{z_2 - z_1}$$

maps as required.
In the general case, one sets

$$M := M_{[w_1, w_2, w_3]}^{-1} \circ M_{[z_1, z_2, z_3]}.$$ □

Two observations play an important role below:

- Transformations with fixed points in $\{0, \infty\}$ are particularly easy to describe.

- One can "rewrite" any Möbius transformation so that the fixed points lie in $\{0, \infty\}$.

The first observation is specified in the next proposition:

Proposition 5.4.4
(i) Suppose that $M = M_{a,b,c,d}$ has $\{0, \infty\}$ as its fixed point set. Then $c = b = 0$, i. e. $M(z) = (a/d)z$, where $a/d \neq 0$. The converse is also true.
(ii) Suppose that $M = M_{a,b,c,d}$ has $\{\infty\}$ as its fixed point set. Then $c = 0$, $a/d = 1$ and $b \neq 0$. The converse is also true.

Proof This follows immediately from proposition 5.4.1.   □

In summary: The M, for which the fixed points are equal to $\{0, \infty\}$ or equal to $\{\infty\}$, are either of the form αz with $\alpha \notin \{0, 1\}$ or of the form $z + \beta$ with $\beta \neq 0$.

5.5   Conjugate Möbius Transformations

We begin with a reminder of linear algebra: Two $n \times n$ -matrices $T, \hat{T}$ can be considered "equivalent" if $\hat{T} = STS^{-1}$ holds for an invertible matrix S: T has the same eigenvalues as $\hat{T}$, T is singular if and only if T' is singular, etc. In general, one considers in a group G conjugate elements x, gxg^{-1} to be "very closely related". Similar to the case of plane motions, we define:

Definition 5.5.1
Two Möbius transformations $M, \hat{M}$ are *conjugate to each other* if there is a Möbius transformation G with $\hat{M} = GMG^{-1}$.

"Conjugate to each other" is obviously an equivalence relation, and for almost all of the properties we are interested in, conjugate transformations both have this property or neither of them. We illustrate this with some examples in the following

Lemma 5.5.2
Let M and $\hat{M} = GMG^{-1}$ be conjugate to each other.

(i)   Let $n \in \mathbb{N}$. Then $M^n = \mathrm{Id}$ holds exactly when $\hat{M}^n = \mathrm{Id}$.
(ii) M commutes exactly when with a mapping N, if $\hat{M}$ commutes with GNG^{-1}.
(iii) Let K be a subset of $\hat{\mathbb{C}}$. Then $M(K) = K$ holds exactly when $\hat{M}(K') = K'$ (with $K' := G(K)$). In particular, $G(z_0)$ is exactly a fixed point for $\hat{M}$, if z_0 is a fixed point for M. It also follows that M and $\hat{M}$ have the same number of fixed points.
(iv) If M and $\hat{M}$ are conjugated, then so are M^{-1} and $\hat{M}^{-1}$ (even under the same mapping G).

Proof (Clear)   □

Proposition 5.5.3
Let $M = M_{a,b,c,d}$ and $G = M_{\alpha,\beta,\gamma,\delta}$ be normalized Möbius transformations and $\hat{M} = GMG^{-1}$. Write $\hat{M}$ in the form $\hat{M}(z) = (Az + B)/(Cz + D)$.
(i) Then one has
$$A = \alpha\delta a - \alpha\beta c + \delta\gamma b - \beta\gamma d,$$
$$B = \beta\delta a + \delta^2 b - \beta^2 c - \beta\delta d,$$
$$C = -\alpha\beta a + \alpha^2 c - \gamma^2 b + \alpha\gamma d,$$
$$D = -\beta\gamma a + \alpha\beta c - \gamma\delta b + \alpha\delta d.$$
(ii) $A + D \in \{-(a + d), a + d\}$.

Proof (i) Two facts need to be combined:

- G^{-1} can be written as $M_{\delta,-\beta,-\gamma,\alpha}$ (Lemma 5.2.2).
- $M_{a',b',c',d'} \circ M_{a'',b'',c'',d''} = M_{a''a'+b''c',a''b'+b''d',c''a'+d''c',c''b'+d''d'}$ (cf. proposition 5.2.3).

This results in the claimed formulas.

(ii) $A + D = \alpha\delta a - \alpha\beta c + \delta\gamma b + \beta\gamma d - \beta\gamma a + \alpha\beta c + \gamma\delta b + \alpha\delta d$
$= a(\alpha\delta - \beta\gamma) + d(\alpha\delta - \beta\gamma)$
$= a + d.$

However, if one has chosen $G^{-1} =$ as $M_{-\delta,\beta,\gamma,-\alpha}$ (an equally valid second possibility), then $A + D = -(a + d)$ follows.   □

5.6   Characterization: Fixed Points in $\{0,\infty\}$

We will show here that, up to equivalence, there are exactly four types of Möbius transformations that are not the identity. In this subsection we focus on the case where the fixed points are in $\{0, \infty\}$. Let $M = M_{a,b,c,d}$ be such a transformation, we want to normalize it. Thus $M(z) = (a/d)z + (b/d)$, where $ad = 1$ holds.

Lemma 5.6.1

Let $a, d \in \mathbb{C}$ with $ad = 1$.

 (i) $a/d = 1$ holds exactly if $a + d \in \{-2, 2\}$
 (ii) $a/d \in\ ]0, \infty[\ \backslash\{1\}$ holds exactly if $a + d \in]-\infty, -2[\cup]2, +\infty[$.
 (iii) Precisely then hold $|a/d| = 1$ and $a/d \neq 1$, if $a + d \in\ ]-2, 2[$.
 (iv) For the number $\alpha := a/d$ it is always true that $\alpha + 1/\alpha = (a + d)^2 - 2$.

Proof (i) $a/d = ad = 1$ implies $a^2 = 1$, thus $a = 1$ or $a = -1$. In the first case, $d = 1$, in the second $d = -1$. Conversely: If, for example, $a + d = 2$, then $a + 1/a = 2$, i. e., $(a - 1)^2 = 0$. This means $a = 1 = d$. Quite analogously, $a + d = -2$ implies that $a = -1 = d$.

(ii) Let $a/d = \lambda$ with $0 < \lambda < +\infty$, $\lambda \neq 1$. Because of $ad = 1$, $a^2 = \lambda$. Thus, $a = \sqrt{\lambda}$ or $a = -\sqrt{\lambda}$, thus $a \neq \pm 1$. In the first case, $a + d = \sqrt{\lambda} + 1/\sqrt{\lambda} > 2$, in the second $a + d = -\sqrt{\lambda} - 1/\sqrt{\lambda} < -2$. (We have used the fact that $x + 1/x$ on $]0, +\infty[$ has the minimum 2, which is only assumed at $x = 1$; this can be easily shown with elementary analysis.)

Conversely: If, for example, $\mu := a + d$ is real and greater than 2, then $a + 1/a = \mu$, and this implies $a = (\mu \pm \sqrt{\mu^2 - 4})/2$. It follows that

$$a/d = a^2 = \frac{\left(\mu \pm \sqrt{\mu^2 - 4}\right)^2}{4} > 0.$$

The case $a + d < -2$ can be handled analogously.

(iii) Write a/d as $e^{i\phi}$ with a $\phi \in\ ]0, 2\pi[$. From $ad = 1$ it then follows that $a^2 = e^{i\phi}$, so $a = \pm e^{i\phi/2}$. This implies that $d = \pm e^{-i\phi/2} = \bar{a}$, and that in turn yields $a + d = 2\operatorname{Re} a \in\ ]-2, 2[$ (because $\phi/2 \in\ ]0, \pi[$).

Conversely, suppose that $a + d = a + 1/a \in\ ]-2, 2[$. Then a is not real, because for real x we have $x + 1/x \geq 2$.

Write a as $re^{i\phi}$. We must have $r = 1$, because otherwise $a + 1/a$ would not be a real number. (The imaginary part is equal to $(r - 1/r) \sin\phi$, and $\sin\phi \neq 0$.) From $a = e^{i\phi}$ it follows that $d = e^{-i\phi}$, so $a/d = e^{2i\phi}$. If $e^{2i\phi} = 1$, then $a = d = -1$ or $a = d = 1$ would follow, so $a + 1/a \notin\ ]-2, 2[$.

(iv) Otherwise stated, the formula says that

$$\frac{a}{d} + \frac{d}{a} = (a+d)^2 - 2,$$

and that follows immediately from $ad = 1$ after multiplication by ad.   □

Now we can list all the possibilities for a Möbius transformation $M = M_{a,b,c,d}$ for which the fixed points lie in $\{0, \infty\}$:

- *Case 1*: M only has the fixed point ∞, so it is of the form $M(z) = z + \beta$ for a $\beta \neq 0$. This is the case exactly when $a/d = 1$ holds, i.e. $a + d \in \{-2, 2\}$. We then call M *parabolic*. (Cf. also Definition 5.7.1.)
- *Case 2*: M has the fixed points $0, \infty$, so it is of the form $M(z) = \alpha z$ for a $\alpha \neq 1$.
  Case 2a: α is real and positive (and of course $\neq 1$). This is due to the above lemma exactly when $a + d \in \;]-\infty, -2[\cup]2, +\infty[$. M is then called *hyperbolic*.
  Case 2b: We have $|\alpha| = 1$ (and also $\neq 1$). This holds exactly when $a + d \in \;]-2, 2[$. M is then called *elliptic*.
  Case 2c: α is neither positive nor has α magnitude one. So $\alpha = re^{i\phi}$, where $r > 0, r \neq 1$ and ϕ is not a multiple of 2π. This is true exactly when $a + d$ is not real. M is then called *loxodromic*.

We now want to put together the essential aspects of these transformations.

The parabolic case $M : z \mapsto z + \beta$ translates a point z by β, where all $\beta \neq 0$ are allowed. The following image indicates this, points at the end of an arrowhead are mapped to the next end of the arrowhead:

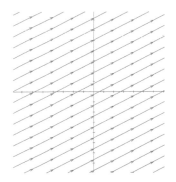

A typical parabolic transformation with fixed point ∞

It is then obvious that the following assertion is true:

Proposiiton 5.6.2
(i)   Invariant (generalized) circles are all lines of the form $z_0 + \mathbb{R}\,\beta$.
(ii) The (generalized) circles orthogonal to the invariant (generalized) circles are all lines of the form $z_0 + \mathbb{R}\,i\beta$.
(iii) $\lim M^n z = \lim M^{-n} z = \infty$ for all z.

The following picture shows some invariant (generalized) circles and the (generalized) circles orthogonal to them:

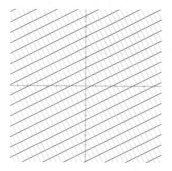

Parabolic: Invariant circles (*dark gray*) and orthogonal circles (*light gray*)

*The hyperbolic case $M : z \mapsto \alpha z$ with an $\alpha \neq 1$ and $\alpha > 0$ only changes the distance to the origin. The effect of this mapping can be imagined as follows:

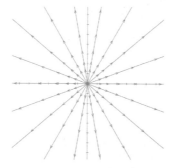

A typical hyperbolic transformation with fixed points 0 and ∞. (In the example, $\alpha > 1$)

In the case of $0 < \alpha < 1$, the distances to 0 would be reduced. The arrowheads would therefore point to the origin. The size of α determines how much stretching or compression is in each case.

It is then obvious that

> **Proposition 5.6.3**
> (i) Invariant (generalized) circles are all lines through the origin, i.e. lines of
>   the form $\mathbb{R}\, z_0$.
> (ii)   The circles orthogonal to the invariant (generalized) circles are the cir-
>   cles with center 0.
> (iii)   If $\alpha > 1$ then $\lim M^n z = \infty$ and $\lim M^{-n} z = 0$ for all $z \neq 0$. For $\alpha < 1$
>   it is the other way around.

In the following picture, some invariant (generalized) circles and the orthogonal
(generalized) circles are shown:

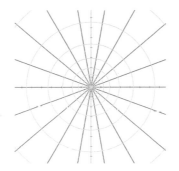

Hyperbolic: Invariant circles (*dark gray*) and orthogonal circles (*light gray*)

*The elliptical case $M : z \mapsto \alpha z$ with an $\alpha \neq 1$ and $|\alpha| = 1$ rotates a point, the point
of rotation is the origin. If $\alpha = e^{i\phi}$, the rotation can go to the left (if $0 < \phi < \pi$),
to the right (if $-\pi < \phi < 0$) or correspond to a reflection (if $\alpha = -1$). The effect
of this mapping can be imagined as follows:*

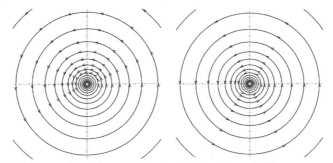

Two typical elliptical transformations with fixed points 0 and ∞ and different α

In the right image there is a special case: After 6 steps, each point is back at itself, so $M^6 = \text{Id}$ holds. In general: If α/π is rational, then $M^k = \text{Id}$ is true for a suitable $k \in \mathbb{N}$.

Then obviously

Proposition 5.6.4
(i)   Invariant circles are all circles with center 0.
(ii) The (generalized) circles orthogonal to the invariant circles are the lines through the origin, i.e. lines of the form $\mathbb{R} z_0$.
(iii) Except for $z = 0$ and $z = \infty$, no sequence $(M^n z)$ is convergent.

In the following image, some invariant (generalized) circles and the orthogonal (generalized) circles are shown:

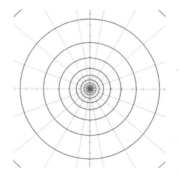

Elliptic: invariant circles (*dark gray*) and orthogonal generalized circles (*light gray*)

The loxodromic case These are maps $M : z \mapsto \alpha z$, where $|\alpha| \neq 1$ and $\alpha \notin \mathbb{R}^+$, i.e. rotations or rotostretches. A typical loxodromic transformation looks like this:

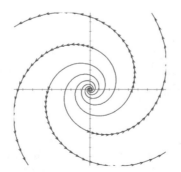

A typical loxodromic transformation with fixed points 0 and ∞ and $|\alpha| > 1$

Depending on whether $|\alpha| < 1$ or > 1 holds, the spirals run outwards or inwards. Then

Proposition 5.6.5
(i) There are no invariant circles (but there are invariant spirals).
(ii)   If $|\alpha| > 1$ then $\lim M^n z = \infty$ and $\lim M^{-n} z = 0$ for all $z \neq 0$. For $|\alpha| < 1$ it is the other way around.

The following image shows some invariant spirals:

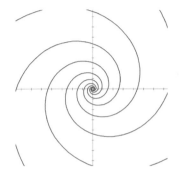

Loxodromic: invariant spirals

Here are *two final remarks*.

1. A Möbius transformation M is invertible, thus $\{M^n \mid n \in \mathbb{Z}\}$ generates a group of maps. Writing $M_n := M^n$, $n \mapsto M_n$ is a group homomorphism defined on $\mathbb{Z}$. Interestingly, it can be extended to a group homomorphism defined on $\mathbb{R}$. M_t is thus defined for all $t \in \mathbb{R}$, and $M^n = M_n$ for all n: The M_t interpolate between the M_n. We illustrate this with the Möbius transformations with fixed points at $\{0, \infty\}$, by conjugation one can extend this to all M.
   Parabolic. M is defined by $M : z \mapsto z + \beta$. Define M_t by $z \mapsto z + t\beta$ for $t \in \mathbb{R}$.
   Hyperbolic. This time $M : z \mapsto \alpha z$ with a real positive α. With $\gamma := \log \alpha$ we set $M_t(z) := e^{t\gamma} z$ for $t \in \mathbb{R}$. (However, there are other possibilities. Choose a $k \in \mathbb{N}$ and set $M_t(z) := e^{t\gamma + 2kt\pi i} z$. Then the orbits circle on spirals towards 0 or ∞.)
   Elliptical. This time $M : z \mapsto \alpha z$ with an α of magnitude 1. Write $\alpha = e^{i\phi}$ and set $M_t(z) := e^{it\phi} z$ for $t \in \mathbb{R}$. Again, there are several reasonable ways to define the M_t because α can also be written as $\alpha = e^{i(\phi + 2k\pi)}$ (for any $k \in \mathbb{Z}$).
   Loxodromic. M is of the form $z \mapsto \alpha z$, where $\alpha = re^{i\phi}$ with $r > 0$, $r \neq 1$, $e^{i\phi} \neq 1$. Write $\rho := \log r$ and set $M_t(z) := e^{t(\rho + i\phi)} z$ for $t \in \mathbb{R}$. In this case too, there are many more possibilities.
   If you imagine these trajectories on the globe, then:
   Parabolic. All trajectories form circles that go through the North Pole.
   Hyperbolic. The trajectories run along the meridians (all from north to south or all the reverse).

Elliptical. The trajectories run on the parallels

Loxodromic. The trajectories run in a spiral: out of the South Pole and into the North Pole or vice versa.

2. We are still investigating how many "really different" Möbius transformations with fixed points in $\{0, \infty\}$ exist. For $\beta \neq 0$ we put $T_\beta(z) := z + \beta$. (A translation, the typical parabolic transformation.) And for $\alpha \neq 0$ we put $M_\alpha: z \mapsto \alpha z$. (These multiplication operators contain the hyperbolic, the elliptic and the loxodromic transformations.)

Proposition 5.6.6

(i)   No T_β is conjugate to any M_α.

(ii) Any two $T_\beta, T_{\beta'}$ are conjugate.

(iii) M_α and $M_{\alpha'}$ are conjugate if and only if $\alpha\alpha' = 1$ or $\alpha = \alpha'$.

Proof (i) Conjugate transformations have the same number of fixed points. T_β has one, the M_α have two or infinitely many. (The result can also be proved with the help of proposition 5.5.3: When conjugating, the trace $a + d$ is retained except for the sign.)

   (ii) Let $G(z) := \gamma z$ with a $\gamma \neq 0$. Then

$$GT_\beta G^{-1}(z) = G(z/\gamma + \beta) = z + \gamma\beta,$$

i.e. $GT_\beta^{-1}G = T_{\gamma\beta}$. Therefore, all $T_\beta, T_{\beta'}$ are conjugated.

   (iii) If $\alpha\alpha' = 1$, then $M_{\alpha'} = M_\alpha^{-1} = M_{1/\alpha}$. These two transformations are conjugated by $G(z) = 1/z$:

$$GM_\alpha G^{-1} = 1/(\alpha/z) = \frac{1}{\alpha}z = M_{\alpha'}(z).$$

To prove the reverse, let $M_\alpha, M_{\alpha'}$ be conjugated. (We will have to show that $\alpha\alpha' = 1$ or $\alpha = \alpha'$.) By assumption there is $G = M_{a,b,c,d}$ with $GM_\alpha = M_{\alpha'}G$, i.e.

$$\frac{a\alpha z + b}{c\alpha z + d} = \alpha'\frac{az + b}{cz + d}$$

for all z.

   Case 1: $c = 0$. We then know that

$$\frac{a\alpha z + b}{d} = \alpha'\frac{az + b}{d}$$

always holds, and $a \neq 0 \neq d$. If $b \neq 0$, then put $z = 0$. It follows that $b/d = \alpha'b/d$, so $\alpha' = 1$. Therefore $GM_\alpha = G$, i.e. $M_\alpha = \mathrm{Id}$, i.e. $\alpha = 1$.

   If, on the other hand, $b = 0$, this means $a\alpha z/d = \alpha'az/d$, and this is only possible for $\alpha = \alpha'$.

Case 2: $c \neq 0$. First, let $a \neq 0$. We let z converge against ∞, it follows $\alpha a/c = \alpha' a/c$, so $\alpha = \alpha'$.

$a = 0$ means that

$$\frac{b}{c\alpha z + d} = \alpha' \frac{b}{cz + d}$$

holds, and we know that $b \neq 0$. In the case $d = 0$, $b/(c\alpha z) = \alpha' b/(cz)$ follows, so $\alpha \alpha' = 1$. In the case $d \neq 0$, we set $z = 0$ and obtain $b/d = \alpha' b/d$, so $\alpha' = 1$. As above, it follows that $\alpha = 1$ is also true. □

Corollary 5.6.7

Let $M, \hat{M}$ be conjugate transformations with fixed point set $\{0, \infty\}$ or $\{\infty\}$. Then both are parabolic, or both are hyperbolic, or both are elliptical or both are loxodromic.[5]

5.7 Characterization: the General Case

We combine the following results:

- We can characterize Möbius transformations if the fixed points lie in $\{0, \infty\}$.
- Every Möbius transformation is conjugate to a Möbius transformation with fixed points in $\{0, \infty\}$
- The number $a + d$ is – except for the factor ± 1 – an invariant under conjugation (proposition 5.5.3).

Definition 5.7.1

Let $M = M_{a,b,c,d}$ be a normalized Möbius transformation that is not the identity. We call M

- parabolic, if $a + d \in \{-2, 2\}$;
- hyperbolic, if $a + d \in \,]-\infty, 2[\,\cup\,]2, +\infty[$;
- elliptic, if $a + d \in \,]-2, 2[$
- loxodromic, if $a + d \notin \mathbb{R}$.

These properties are *well-defined*, because it does not matter whether one defines M as $M_{a,b,c,d}$ or as $M_{-a,-b,-c,-d}$. The names also agree with those of the previous section, if the fixed points of M are in $\{0, \infty\}$. It is clear that each M different from the identity belongs to exactly one of the four classes.

---

[5] This result is valid for any Möbius transformation: see below, Corollary 5.7.2.

Proposition 5.5.3 has the following consequence:

Corollary 5.7.2

Let M and $\hat{M}$ be conjugate transformations. Then M is exactly parabolic or hyperbolic or elliptic or loxodromic if $\hat{M}$ is parabolic or hyperbolic or elliptic or loxodromic.

In order to better understand these special Möbius transformations, we first look for a Möbius transformation $\hat{M}$ conjugate to M, which has its fixed points in $\{0, \infty\}$:

Case 1: M has only one fixed point ξ. Choose z_1, z_2, so that z_1, z_2, ξ are pairwise different. Then determine G with the property $z_1, z_2, \xi \mapsto 0, 1, \infty$ (proposition 5.4.3). $\hat{M} := GMG^{-1}$ then has the only fixed point ∞.

Case 2: M has the two fixed points ξ_1, ξ_2. Choose z_1 so that z_1, ξ_1, ξ_2 are pairwise different. Then determine G with the property $z_1, \xi_1, \xi_2 \mapsto 1, 0, \infty$. $\hat{M} := GMG^{-1}$ then has the two fixed points $0, \infty$.

Then we already know how $\hat{M}$ looks. There are *the following possibilities*[6] :

$\hat{M}$ *is parabolic, that is, of the form* T_β Consequently, $M = G^{-1}T_\beta G$ for a suitable G. Typically, M can then be visualized by transforming the images of the previous section with G^{-1}. Here it is done in reverse. A Möbius transformation G was chosen and then M was defined by $GT_\beta G^{-1}$. The images under G of the invariant subsets (parallel lines) and the orthogonal "circles" (also parallel, orthogonal to the first lines) are shown.

First, however, one finds a colored illustration of the case where ∞ is the only fixed point. The map translates to the upper right, which is represented by three points z, Mz, M^2z (black, gray, gray):

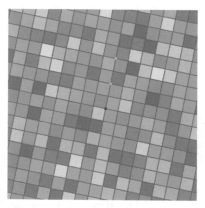

A parabolic transformation with the only fixed point ∞ (with colors "decorated")

By transformation you get a typical image if there is only one finite fixed point. You just have to map the lines under G, which will generally be circles:

---

[6] For the notations see the previous section.

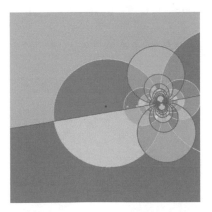

A typical parabolic transformation with a single finite fixed point (*colored*)

One sees:

- Some invariant curves for M (dark gray). These are in this case circles with the exception of one straight line. The exception is a line through the fixed point.
- The circles orthogonal to the invariant curves (light gray).
- The action of M (qualitatively): The black point is mapped under M to the next gray and then to another gray.

One can imagine the action of M as a large whirlpool that drives particles on circles that go through the fixed point. The direction of rotation is clockwise on the upper circles and counterclockwise on the lower circles (and on the only line from left to right.) Or vice versa.

However, the whole truth is a little more complicated, the dynamics on the line that is touched by all fixed circles is somewhat complicated. As a typical example of a parabolic transformation, we consider the transformation $M(z) := z/(z + 1)$. The origin is a double fixed point, and $\mathbb{R}$ is an invariant line. So typically the z move under M on circles whose centers are on $i\mathbb{R}$ and that touch $\mathbb{R}$ at 0 tangentially. All movements on the circles take place clockwise, the dynamics on the invariant line $\mathbb{R}$ corresponds to $M : t \mapsto t/(1 + t)$. Here is a table:

t	-11	$-1{,}1$	-1	-0.5	-0.1	0	$0{,}1$	1	2	100
$M(t)$	$1{,}1$	11	∞	-1	$-1/9$	0	$1/11$	0.5	2/3	100/101

If you imagine $\mathbb{R}$ as a circle (below is zero, above is ∞), then the following happens:

- All points are rotated clockwise, the zero is fixed.
- If P is further clockwise than Q, then $M(P)$ is further than $M(Q)$.

- The most remarkable jump is at -1, where it goes directly to the point ∞.

$\hat{M}$ *is hyperbolic, thus of the form* M_α *for an* $\alpha \in \,]0, +\infty[\setminus \{1\}$ Typically, you can imagine M like this:

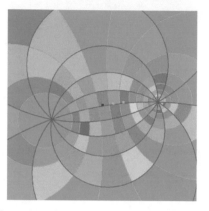

A typical hyperbolic transformation with two finite fixed points (*colored*)

One sees:

- The invariant subsets (dark gray).
- The orthogonal circles (light gray).
- The effect of M (black-gray-gray): Points are "sucked in" from the right fixed point to circle arcs.

M thus acts as if there were a powerful source and a powerful sink that attracts everything on circle arcs from the source. Interesting is the effect of M on the connecting line between the fixed points: If the point is between the fixed points, it moves straight from one fixed point to the other, otherwise it takes the detour via ∞.

$\hat{M}$ *is elliptical, so of the form* M_α *for a* $\alpha \neq 1$ *with* $|\alpha| = 1$ Typically one can imagine M as follows:

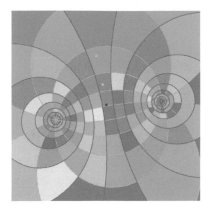

A typical elliptical transformation with two finite fixed points (*colored*)

One sees:

- The invariant subsets (dark gray).
- The orthogonal circles (light gray).
- The effect of *M* (black-gray-gray): Points move on circular arcs around the fixed points.

So there are two "vortices", one around each fixed point. Around one they move counterclockwise, around the other clockwise. (And quite complicated on the line that is orthogonal to the connecting line of the fixed points; see the corresponding discussion for the parabolic case.) By the way, this line cuts the connecting line of the fixed points orthogonally in the middle (see the next corollary).

$\hat{M}$ *is loxodromic, that is, of the form* M_α *with* $\alpha \notin \mathbb{R}^+, |\alpha| \neq 1$ *These transformations lead to the most interesting images. Here are two examples:*

Two typical loxodromic transformation to different α and two finite fixed points (*colored*)

One sees:

- The invariant subsets (dark gray).
- The transformed images of some circles (light gray).
- The effect of M (black-gray-gray): Points move on circular arcs around the fixed points.

There are thus two "vortices", one around each fixed point. The left fixed point acts in the examples as a "source", the right one as a "sink". Depending on α, the vortices run to the left or right, and they can also be of different strengths.

A mapping $z \mapsto \alpha z$ has no orthogonal circles in the loxodromic case. Nevertheless, all circles $|z| = r$ are cut by the invariant curves $t \mapsto z_0 \exp t(\rho + i\gamma)$ at the same angle:

At time t, the point $P := z_0 \exp t(\rho + i\gamma)$ will be reached. It lies on the circle with radius $|P|$, the direction of the circle tangent is iP. The curve itself has the velocity vector $(\rho + i\gamma)P$ there. Note that the angle between two complex numbers z, w is equal to ϕ if z/w is written as $re^{i\phi}$ with real ϕ. Here it is about the quotient $(\rho + i\gamma)/i = \gamma - i\rho$, and that is independent of P.

This is then also true for the transformed circles and curves due to proposition 5.3.3.

We have seen that for the qualitative behavior of a normalized Möbius transformation $M = M_{a,b,c,d}$, only the type and, in the case of a non-parabolic transformation, also the number α are of importance. α is however not uniquely determined, instead of α one can also consider $1/\alpha$. (Which of these numbers occurs depends on how the fixed points are numbered). α is called the *multiplier* of M. We claim:

Lemma 5.7.3

(i) If one chooses an $\hat{\alpha}$ with $\hat{\alpha}^2 = \alpha$, then

$$\hat{\alpha} + \frac{1}{\hat{\alpha}} \in \{a + d, -(a + d)\}.$$

applies

(ii) $\alpha + 1/\alpha = (a + d)^2 - 2$.

(iii) Suppose that M has two different fixed points. Then

$$\frac{M(z) - \xi_1}{M(z) - \xi_2} = \alpha \frac{z - \xi_1}{z - \xi_2},$$

applies, where ξ_1, ξ_2 denotes the fixed points of M.

Proof (i) M has – possibly up to the sign – the same $a + d$ as the transformation αz, since both transformations are conjugated. (Cf. proposition 5.5.3.) However, it must be normalized, for example as $\hat{\alpha}/(1/\hat{\alpha})$, so that it is correct. The claim follows.

(ii) This follows by squaring the formula in (i).

(iii) Let $G(z) = (z - \xi_1)/(z - \xi_2)$. This Möbius transformation maps ξ_1 to 0 and ξ_2 to ∞, and now we conjugate: GMG^{-1} has the fixed points 0 and ∞ and is therefore of the form M_α. We have $GM = M_\alpha G$, and this is a reformulation of the claim.   □

Corollary 5.7.4
Let M be elliptical. Then the line that is perpendicular to the line segment from ξ_1 to ξ_2 and passes through the midpoint of this segment is an invariant set for M.

Proof The relevant line is given by $\{z \mid |z - \xi_1| = |z - \xi_2|\}$, and with z, because of $|\alpha| = 1$, also $M(z)$ has the property that the distances to ξ_1 and ξ_2 are equal.   □

5.8   Wish List/Visualization

Wish list Ideas successfully used in the previous section enable us to construct "custom-made" Möbius transformations.

Parabolic: There is only one fixed point z_0. If it is equal to ∞, then $z \mapsto z + \beta$ can be specified with any $\beta \neq 0$. If z_0 is finite, one still needs the information in which direction the invariant line from z_0 to ∞ is supposed to run. Let's assume it should run in the direction of w_0, so it goes through z_0 and $z_0 + w_0$.

$G(z) := 1/(z - z_0)$ maps z_0 to ∞ and ∞ to 0. The line through z_0 and $z_0 + w_0$ and ∞ should therefore become the line from 0 to ∞. This is the line $\beta\mathbb{R}$. So it is enough, to choose β such that $G(z_0 + w_0) = 1/w_0 = \beta$. In short: The desired Möbius transformation is $G^{-1}T_{1/w_0}G$.

Non-parabolic: We focus on the case that both fixed points, ξ_1 and ξ_2, are finite. So you want to prescribe freely ξ_1 and ξ_2 and the multiplier $\alpha \in \mathbb{C} \setminus \{1\}$. $G(z) = (z - \xi_1)/(z - \xi_2)$ maps ξ_1 and ξ_2 to 0 and ∞, and GMG^{-1} (with a still unknown M) should be M_α. It follows that $M = G^{-1}M_\alpha G$ has the desired properties.

Visualization It is helpful to visualize the effect of a Möbius transformation M. One proceeds as follows:

- M has to be normalized, that is, ensure $ad - bc = 1$. From $a + d$ you can then read the type.
- Determine the fixed points ξ_1 and ξ_2.

Case 1: There is only one fixed point ξ_0. Transform to GMG^{-1}, where G is the mapping $G(z) = 1/(z - \xi_0)$. This must be a transformation T_β, because the only fixed point is ∞. Draw "many" lines $z_0 + \mathbb{R}\beta$, these are examples of invariant subsets of T_β. Transform the lines back by mapping a line K to the line or circle $G(K)$. This results in the typical parabolic pattern in which many circles touch tangentially at a single point (the only fixed point).

Below you will find a visualization of $M(z) = z/(2z + 1)$ (0 is a double fixed point) and $M(z) = 3z/(4z - 1)$ (0.5 is a double fixed point):

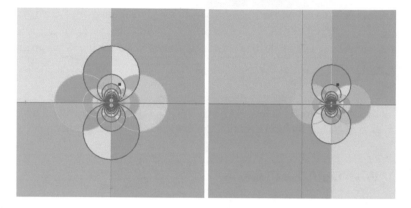

Two typical parabolic transformations

Case 2: There are two different fixed points. Transform them so that the fixed points are 0 and ∞ and read the multiplier α from $a + d$ ($\alpha + 1/\alpha = (a + d)^2 - 2$). Draw circles in $\mathbb{C}$ and lines emanating from 0 (or, in the loxodromic case, invariant spirals) and transform them back. Draw the sequence z, Mz, M^2z for one or more z. Here are some examples:

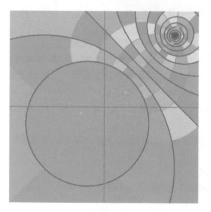

The elliptical transformation $((-1 + i)z + 1)/(0.5iz + 1 - i)$

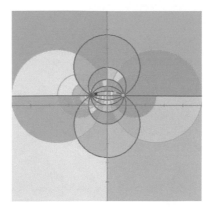

The hyperbolic transformation $\big((2+i)z+2\big)/(2z+2-i)$

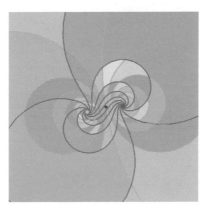

The loxodromic transformation
$$\big((1.018+0.005i)z+0.123+0.169i\big)/\big((0.132+0.091i)z+0.983+0.027i\big)$$

Groups of Möbius Transformations

6

The set of all Möbius transformations forms a group. In some sense, they play the same role for $\mathbb{C}$ as the movements of the plane for $\mathbb{R}^2$ from the first part.

We will be interested in subgroups with special properties. As in Euclidean theory, the elements "cannot come arbitrarily close". More precisely: If $\mathcal{G}$ is a group of Möbius transformations, then $\mathcal{G}$ is called *discontinuous* (or also *discrete*), if there is a z_0 and an environment U of z_0 such that $Mz_0 \notin U$ for all $M \in \mathcal{G} \setminus \{\mathrm{Id}\}$. One then also speaks of a *Kleinian group*.

Such groups are named after *Felix Klein* (1849–1925), who intensively studied groups of this type. In contrast to the Euclidean case, they cannot simply be characterized. Here we will study some special cases of Kleinian groups that have proven to be particularly interesting.

The historical starting point were special questions from the theory of complex analytic functions. The search for the understanding of multivalued functions (such as roots or logarithms) leads to the question of how meromorphic functions f can look, which are defined on a domain $G \subset \mathbb{C}$ and for which $f(z) = f\big((M(z))\big)$ for all M from a certain group $\mathcal{G}$ of Möbius transformations applies. If there were a $z \in G$, so that $\{Mz \mid M \in \mathcal{G}\}$ has an accumulation point w in G, then f would be canstant with value $f(w)$. (Namely, if $f(w) = f(w_n)$ holds for a sequence w converging to (w_n), then f must be constant; see Sect. 5.1.) To exclude such trivial cases, one restricts oneself to the study of discrete groups.

6.1   First Examples of Groups of Möbius Transformations

The set of all Möbius transformations forms a group. We are interested in "small" subgroups, which often already give rise to quite complicated problems. And sometimes our results can be translated into attractive images.

Here are some examples:

1. Fix a Möbius transformation M and consider

$$\mathcal{G}_M = \{M^n \mid n \in \mathbb{Z}\},$$

   the *cyclic group* generated by M. It can only be finite in the elliptic case.
2. Fix $\Delta \subset \hat{\mathbb{C}}$ and define

$$\mathcal{G}_\Delta = \{M \mid M(\Delta) = \Delta\},$$

   which is the set of transformations M that map Δ to itself. The special cases
   and will be studied in more detail in the next section.
   For one-pointed Δ it is about Möbius transformations that all have the same
   fixed point.
3. If $\mathcal{M}$ is a set of Möbius transformations, then $\mathcal{G}(\mathcal{M})$ denotes the subgroup generated by $\mathcal{M}$. Example 1 concerns the case where $\mathcal{M}$ consists of only one transformation, but for two-element $\mathcal{M}$ a very complicated structure can arise.
   Here is a simple concrete example. $z_0, w_0 \in \mathbb{C}$ are given, these numbers should
   be $\mathbb{R}$-linearly independent. (It should *not* be one a real multiple of the other.)
   M_1 or M_2 is the transformation $z \mapsto z + z_0$ or $z \mapsto z + w_0$. If $\mathcal{G}_{z_0,w_0}$ is the subgroup generated by M_1, M_2, then it is easy to see that $\mathcal{G}_{z_0,w_0}$ consists exactly of
   the transformations $z \mapsto z + nz_0 + mw_0$ with $n, m \in \mathbb{Z}$.
   (In general: If M_1, M_2 commute, then the subgroup generated by M_1, M_2 consists exactly of the $M_1^n M_2^m$ with $n, m \in \mathbb{Z}$.)

As in the first chapter, it is also here the case that one can conjugate subgroups
and that the conjugated subgroup can be considered "essentially equal" to the subgroup from which it started. More precisely: If $\mathcal{G}$ is a subgroup of the group of
Möbius transformations and G is a fixed Möbius transformation, then define

$$G\mathcal{G}G^{-1} := \{GMG^{-1} \mid M \in \mathcal{G}\}.$$

For example, we then have:

- $\mathcal{G}$ is finite exactly when $G\mathcal{G}G^{-1}$ is finite.
- $\mathcal{G}$ consists only of elliptical transformations, if and only if $G\mathcal{G}G^{-1}$ has this property.
- $\mathcal{G}$ is discrete exactly when $G\mathcal{G}G^{-1}$ is discrete.
- $\mathcal{G}$ is commutative exactly when $G\mathcal{G}G^{-1}$ is commutative.
- If all $M \in \mathcal{G}$ leave a set $\Delta \subset \hat{\mathbb{C}}$ invariant, then all GMG^{-1} leave the set $G\Delta$
  invariant.
- ...

6.2 Fundamental Domains and Discrete Groups

Let G be a group of Möbius transformations. Similar to the case of movements in the first chapter, we would like to find "simple" domains $\mathbb{F} \subset \hat{\mathbb{C}}$ so that the sets $M(\mathbb{F})$ for $M \in G$ cover $\hat{\mathbb{C}}$ without overlaps and fill this set completely. Such an should then be called a *fundamental domain*. Even without a completely exact definition, it is then obviously true:

- With $\mathbb{F}$, every $M(\mathbb{F})$ for $M \in G$ is a fundamental domain.
- If $\mathbb{F}$ is a fundamental domain for G and one conjugates G with a transformation G (i.e. passes to $\hat{G} = \{GTG^{-1} \mid T \in G\}$), then $G^{-1}(\mathbb{F})$ is a fundamental domain for $\hat{G}$. (Because $z \in GTG^{-1}(\mathbb{F})$ is equivalent to $G^{-1}z \in T(G^{-1}(\mathbb{F}))$.)

We consider some examples:

1. Let $T(z) := \alpha z$ with $|\alpha| \neq 1$ and G_T be the cyclic group generated by T; T is therefore hyperbolic or loxodromic. For each $r > 0$, the circular ring $\mathbb{F}$ of all z for which $|z|$ lies between r and $|\alpha|r$ is a fundamental domain. In the following picture you can see an example marked in red together with some $T^n(\mathbb{F})$ marked in green. It is also shown how $\mathbb{F}$ looks like after a conjugation.

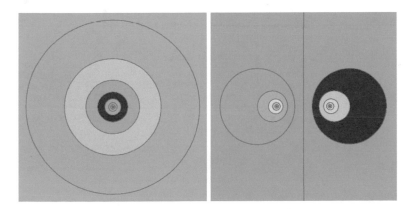

A fundamental domain for G (generated by $z \mapsto \alpha z$) and a conjugate group

2. Let $T(z) := z + \beta$ with $\beta \neq 0$; T is therefore parabolic. Choose two parallel lines in $\mathbb{C}$ so that one is transferred to the other under T and denote as $\mathbb{F}$ the points on or between these lines. This is a fundamental domain for G_T.

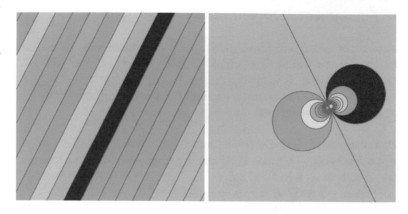

A fundamental domain for $\mathcal{G}$ (parabolic cyclic) and a conjugate group

3. Now let $Tz = \alpha z$ be with a α on the boundary of the unit circle: T is elliptical. "Reasonable" fundamental domains exist exactly when $\alpha = e^{i\phi}$ as an argument has a rational multiple of π, that is, when $T^n = \mathrm{Id}$ holds for a suitable $n \in \mathbb{N}$. In the following example, $\alpha^8 = 1$.

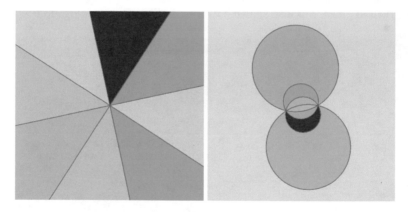

A fundamental domain for $\mathcal{G}$ (elliptic cyclic) and a conjugate group

4. Next we consider the group consisting of all translations $T_{n\beta_1 + m\beta_2}$ $(m, n \in \mathbb{Z})$, where β_1, β_2 are linearly independent over $\mathbb{R}$. As a fundamental domain one can then choose a parallelogram generated by β_1, β_2:

$$\mathbb{F} = \{s\beta_1 + t\beta_2 \mid 0 \leq s, t \leq 1\}.$$

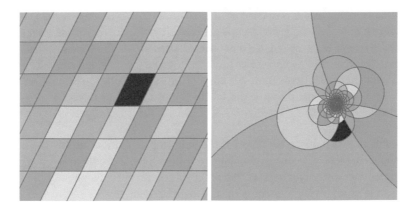

A fundamental domain for $\mathcal{G}$ (2 parabolic generators) and a conjugate group

5. The four-element group $\{\pm z, \pm 1/z\}$ is isomorphic to the Kleinian four-group. Here is a fundamental domain:

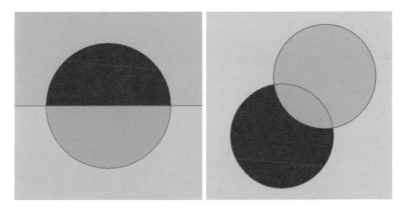

A fundamental domain for $\mathcal{G}$ (Kleinian four-group) and a conjugate group

It should be emphasized that in each of these examples one can also define much more complicated fundamental domains.

But what is a fundamental domain if one wants to define this term precisely?

First Attempt $\mathbb{F}$ is called a fundamental domain, if $\hat{\mathbb{C}}$ is the disjoint union of the $T(\mathbb{F})$, $T \in \mathcal{G}$. Such a set always exists: Call z, w equivalent, if there is a T with $Tz = w$ and choose with the Zorn's lemma $\mathbb{F}$ so that from each equivalence class exactly one element is contained. These $\mathbb{F}$ will usually look very chaotic, often not even measurable.

Second Attempt $\mathbb{F}$ is called a fundamental domain, if $\mathbb{F}$ is open and the $M(\mathbb{F})$ are disjoint and come arbitrarily close to each $z \in \hat{\mathbb{C}}$. In the above examples, one could choose the interior of the proposed $\mathbb{F}$ in each case. However, then some $\mathcal{G}$ have no fundamental domains (like most groups generated by elliptic transformations.)

Third Attempt $\mathbb{F}$ is called a fundamental domain, if $\mathbb{F}$ is closed, $\mathbb{F} = (\mathbb{F}^o)^-$ applies[1], different $M(\mathbb{F})$ intersect at most in boundary points and the $M(\mathbb{F})$ fill the set $\hat{\mathbb{C}}$. With this definition, there are already problems with hyperbolic and loxodromic transformations, because the origin is not contained in any $T^n(\mathbb{F})$ for our fundamental domains.

In summary: You have to be very careful with the definition, and it is not to be expected that there are always fundamental domains. We will take up the topic again later, when we want to find "as good as possible" fundamental domains for any discrete group of Möbius transformations. (Cf. Sect. 6.9.3.)

6.3   Special Möbius Transformations

Sometimes it is interesting to know which Möbius transformations leave special subsets of $\mathbb{C}$ invariant. The following cases are particularly important:

- $\mathbb{U} = \{z \mid |z| < 1\}$. Which M leave the unit circle invariant?
- $\mathbb{H} := \{z \mid \operatorname{Im} z > 0\}$. Which M leave the upper half-plane invariant?[2]

The unit circle Let K be the boundary of the unit circle, that is $K = \{z \mid z\bar{z} - 1 = 0\}$. We claim:

> **Lemma 6.3.1**
> For $A, B, C \in \mathbb{C}$ the set $\{z \mid Az\bar{z} + Bz + \bar{B}\bar{z} + C = 0\}$ equals K exactly when $A = -C \neq 0$ and $B = 0$.

Proof One direction of the proof is clear. Conversely, let A, B, C be such that these numbers describe the unit circle. If we set $z = 1$ and $z = -1$, then $A + B + \bar{B} + C = A - B - \bar{B} + C = 0$ follows, so $B + \bar{B} = 0$: B is purely imaginary. Now we consider i and $-i$, which gives $-A + i(B - \bar{B}) + C = -A - i(B - \bar{B}) + C = 0$. So $B - \bar{B} = 0$ is also true, and that proves $B = 0$.

We consider 1 once more and get $A = -C$. $A = 0$ is not possible here, otherwise the whole plane would be defined by the equation. If we divide by A, the equation $z\bar{z} - 1 = 0$ results. $\square$

---

[1] $\mathbb{F}$ is thus the closure of the interior of $\mathbb{F}$; such sets are called regularly closed.

[2] It is also easy to see by using a continuity argument: If $M(\mathbb{U}) = \mathbb{U}$, then $|M(z)| = 1$ for every z with $|z| = 1$. And from $M(\mathbb{H}) = \mathbb{H}$ it follows that $M(\mathbb{R} \cup \{\infty\}) = \mathbb{R} \cup \{\infty\}$.

Proposition 6.3.2

(i) For a normalized $M = M_{a,b,c,d}$ one has $M(K) = K$ and $M(\mathbb{U}) = \mathbb{U}$, exactly when $b = \overline{c}$ and $d = \overline{a}$ and $a\overline{a} - c\overline{c} = 1$, i.e., when M has the form

$$M(z) = \frac{az + \overline{c}}{cz + \overline{a}}$$

with $|a|^2 - |c|^2 = 1$.

(ii) For a normalized $M = M_{a,b,c,d}$ one has $M(K) = K$ and $M(\mathbb{U}) = \{|z| > 1\}$, precisely when $b = -\overline{c}$ and $d = -\overline{a}$ and $a\overline{a} - c\overline{c} = -1$, i.e., when M has the form

$$M(z) = \frac{az - \overline{c}}{cz - \overline{a}}$$

with $|c|^2 - |a|^2 = 1$.

Proof (i) Assume that $M(z) = \frac{az+\overline{c}}{cz+\overline{a}}$ with $|a|^2 - |c|^2 = 1$. M therefore is normalized. Now let z be arbitrary with $z \in K$. Then

$$M(z)\overline{M(z)} = \frac{az + \overline{c}}{cz + \overline{a}} \cdot \frac{\overline{a}\overline{z} + c}{\overline{c}\overline{z} + a}$$

$$= \frac{|az + \overline{c}|^2}{|cz + \overline{a}|^2}.$$

One has $az + \overline{c} = z(a + \overline{c}\overline{z})$, and therefore

$$|az + \overline{c}| = |a + \overline{c}\overline{z}|$$
$$= |\overline{a + \overline{c}\overline{z}}|$$
$$= |\overline{a} + cz|.$$

This implies $M(z)\overline{M(z)} = 1$, M therefore maps K to K. Because of Lemma 5.2.2 is $M^{-1} = M_{\overline{a},-\overline{c},-c,a}$, which is a transformation with the same structure (one only has to replace a by $\overline{a}$ and c by $-c$). Consequently, $M^{-1}(K) \subset K$, and that proves $M(K) = K$. Also $|M(0)| = |c/a|$, and this number is – due to $|a|^2 - |c|^2 = 1$ – smaller than one: $\{|z| \leq 1\}$ is therefore mapped to itself under M.

For the proof of the converse, we assume that $M(K) = K$ holds, where $M = M_{a,b,c,d}$ is normalized. We know that $M^{-1} = M_{d,-b,-c,a}$. The set $\{z \mid z\overline{z} = 1\}$ therefore equals $\{z \mid M^{-1}(z)\overline{M^{-1}(z)} = 1\}$, and this means

$$\frac{dz - b}{-cz + a} \cdot \frac{\overline{dz - b}}{\overline{-cz + a}} = 1$$

or

$$(dz - b)(\overline{dz - b}) = (-cz + a)(\overline{-cz + a})$$

or

$$Az\overline{z} + Bz + \overline{B}\overline{z} + C = 0$$

with $A = d\overline{d} - c\overline{c}$, $B = -\overline{b}d + c\overline{a}$ and $C = b\overline{b} - a\overline{a}$. Lemma 6.3.1 then guarantees $A = -C \neq 0 = B$.

Case 1: c = 0 or d = 0. Then M would be essentially of the form $z \mapsto 1/z$ or $z \mapsto z + \beta$. That can't be if M is supposed to map $\mathbb{U}$ into itself.

Case 2: c, d \neq 0. $B = 0$ implies $\overline{b}/c = \overline{a}/d =: \lambda$. If you plug that into $A = -C$, you get

$$d\overline{d} - c\overline{c} = \lambda\overline{\lambda}(d\overline{d} - c\overline{c}),$$

so $|\lambda| = 1$. Also, $1 = ad - bc = \overline{\lambda}(d\overline{d} - c\overline{c})$, λ must be real, and that implies $\lambda \in \{-1, 1\}$. If $\lambda = -1$, then $d\overline{d} - c\overline{c} = -1$ would follow, and the singularity $-d/c$ (which is mapped to ∞) would be in $\{|z| \leq 1\}$. That contradicts the assumption, and so $\lambda = 1$. It follows that $b = \overline{c}$ and $d = \overline{a}$.

(ii) In this case we are led to $\lambda = -1$, it follows that $b = -\overline{c}$ and $d = -\overline{a}$.   □

We now know all M with $M(\mathbb{U}) = \mathbb{U}$, they obviously form a group. Interesting consequences arise from the concrete description of these M:

Proposition 6.3.3

(i)   Let M be a Möbius transformation with $M(\mathbb{U}) = \mathbb{U}$. Then M is not loxodromic: Only parabolic, elliptical and hyperbolic transformations are possible.

(ii) If such an M is parabolic, then the fixed point is on the boundary of $\mathbb{U}$.

(iii) If M is hyperbolic, then both fixed points are on the boundary of $\mathbb{U}$.

(iv) If M is elliptical, then one fixed point is inside of $\mathbb{U}$ and the other is outside of $\mathbb{U}$. The boundary of $\mathbb{U}$ is a fixed circle for M.

Proof (i) This follows immediately from $a + d = a + \overline{a} = 2\operatorname{Re}a \in \mathbb{R}$. It is also clear heuristically if one has the images of the different types in mind.

(ii) One has $a + \overline{a} = 2$ (w.l.o.g.), thus $\operatorname{Re}a = 1$. If $c = 0$, then this is only possible for $a = 1$ and M would be the identity. So we can assume $c \neq 0$.

The fixed point is given by

$$\xi = \frac{a-d}{2c} = \frac{\operatorname{Im} a}{c}$$

For the absolute value follows

$$|\xi|^2 = \frac{(\operatorname{Im} a)^2}{|c|^2} = \frac{|a|^2 - (\operatorname{Re} a)^2}{|c|^2} = \frac{|a|^2 - 1}{|c|^2} = 1.$$

(iii) We will assume that $c \in \mathbb{R}^+$ holds. (We can do this by conjugating with $z \mapsto e^{i\phi} z$, a transformation that leaves U invariant.)

Consider for example ξ_1, where we use the formulas for the fixed points:

$$\xi_1 = \frac{i \cdot 2 \operatorname{Im} a + \sqrt{(2 \operatorname{Re} a)^2 - 4}}{2c},$$

where the expression under the square root is positive. Consequently,

$$|\xi_1|^2 = \frac{1}{4c^2} \left(4(\operatorname{Im} a)^2 + 4(\operatorname{Re} a)^2 - 4 \right) = \frac{1}{c^2}(|a|^2 - 1) = 1.$$

(iv) We have

$$\xi_{1,2} = \frac{a - d \pm \sqrt{(d+a)^2 - 4}}{2c}.$$

Consequently,

$$|\xi_1 \xi_2| = \left| \frac{(a - \bar{a})^2 - \left((a + \bar{a})^2 - 4\right)}{4c^2} \right|$$

$$= \left| \frac{1 - a\bar{a}}{c^2} \right|$$

$$= 1.$$

Since M is elliptic, $a + d = 2 \operatorname{Re} a \in\,]-2, 2[$, i. e. $(a+d)^2 - 4$ is a negative real number $-\lambda$. Consequently, the numerator in the formula for $\xi_{1,2}$ is equal to $i(2 \operatorname{Im} a \pm \sqrt{\lambda})$. If ξ_1, ξ_2 had both magnitude one, then $\operatorname{Im} a = 0$ as well as $|c| = \sqrt{1 - (\operatorname{Re} a)^2}$, i. e. $1 - (\operatorname{Re} a)^2 = 1 - |a|^2 = |c|^2$. Also $1 - |a|^2 = -|c|^2$ holds, and this implies $c = 0$, i.e., M is the identity. $\qquad \square$

And so you can imagine these transformations:

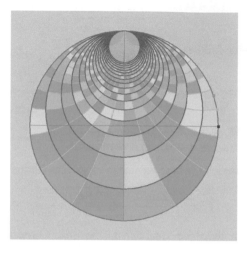

A typical parabolic transformation that leaves $\mathbb{U}$ invariant

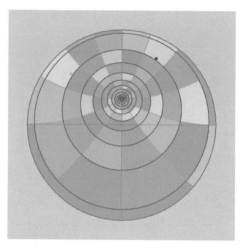

A typical elliptical transformation that leaves $\mathbb{U}$ invariant

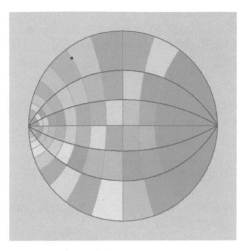

A typical hyperbolic transformation that leaves $\mathbb{U}$ invariant

The upper half-plane $\mathbb{H}$ Which M leave $\mathbb{H}$ invariant? If you recall the effect of the different types of transformations, you guess:

- Loxodromic transformations are not allowed.
- Parabolic transformations are possible if the fixed point is in $\mathbb{R}$ and the tangent line to the circles is equal to $\mathbb{R}$.
- For elliptic transformations, the fixed points must be $\xi_1 = \overline{\xi_2}$.
- The hyperbolic transformations must have both fixed points in $\mathbb{R}$.

That's almost the whole truth. First we prove:

Proposition 6.3.4
The normalized Möbius transformation $M_{a,b,c,d}$ leaves $\mathbb{H}$ invariant if and only if $a, b, c, d \in \mathbb{R}$.

Proof One   direction   is   easy: If   all   coefficients   are   real,   then $M(\mathbb{R} \cup \{\infty\}) = \mathbb{R} \cup \{\infty\}$, and i is mapped to

$$\frac{ai + b}{ci + d} = \frac{(ai + b)(-ci + d)}{c^2 + d^2} = \frac{ca + bd + i(ad - cb)}{c^2 + d^2} = \frac{ca + bd + i}{c^2 + d^2},$$

a number with a positive imaginary part. For continuity reasons, therefore, $M(\mathbb{H}) = \mathbb{H}$.

For the proof of the converse, it is advantageous to transform $\mathbb{H}$ into $\mathbb{U}$. This is accomplished by the *Cayley transformation* $C(z) := (z - i)/(z + i)$. With this definition, $C(\mathbb{H}) = \mathbb{U}$. C maps the points $0, -1, +1$ to points from $\{z \mid |z| = 1\}$, the image of $\mathbb{R}$ is therefore the unit circle; and i is mapped to 0. This shows that $C(\mathbb{H}) = \mathbb{U}$ holds.

To represent C normalized, we choose $\eta \in C$ with $\eta^2 = 1/(2i)$; then

$$Cz = \frac{\eta z - \eta i}{\eta z + \eta i}$$

in normalized representation.

Let a normalized $M = M_{a,b,c,d}$ with $M(\mathbb{H}) = \mathbb{H}$ be given. It has to be shown that a, b, c, d are real. We conjugate M with C, so $\hat{M} = CMC^{-1}$ is a transformation that leaves $\mathbb{U}$ invariant. We write it with suitable A, B, C, D in the form $(Az + B)/(Cz + D)$. It then follows from proposition 6.3.2 that $D = \overline{A}$ and $B = \overline{C}$.

But we can determine A, B, C, D explicitly with proposition 5.5.3. There one puts $\alpha = \eta$, $\beta = -\eta i$, $\gamma = \eta$ and $\delta = \eta i$. A small calculation, in which several times $\eta^2 = -0{,}5i$ is used, results in

$$A = -0{,}5(a + ib - ic + d),$$
$$B = -0{,}5(a - ib - ib - d),$$
$$C = -0{,}5(a + ib + ic - d),$$
$$D = -0{,}5(a - ib + ic + d).$$

The conditions $A = \overline{D}, B = \overline{C}$ yield

$$a + ib - ic + d = \overline{a} + i\overline{b} - i\overline{c} + \overline{d},$$
$$a - ib - ib - d = \overline{a} + -\overline{b} - i\overline{c} - \overline{d}.$$

If you add these equations to each other or subtract them from each other and then divide by 2, you get $a - ic = \overline{a} - i\overline{c}$ and $d + ib = \overline{d} + i\overline{b}$. For real a, b, c, d this is certainly fulfilled, but—surprisingly—only then.

From the first equation it follows $a - \overline{a} = 2i \operatorname{Im} a = i(c - \overline{c}) = -2 \operatorname{Im} c$. But for real x, y $x = iy$ can only then apply if $x = y = 0$, and therefore $\operatorname{Im} a = \operatorname{Im} c = 0$ must hold. Similarly one shows that $b, d \in \mathbb{R}$. $\square$

It follows

Proposition 6.3.5

Let $M = M_{a,b,c,d}$ be a normalized transformation that leaves $\mathbb{H}$ invariant (the a, b, c, d are therefore real). The following cases are possible if M is not the identical mapping:

(i) $a + d \in \{-2, 2\}$, and $c = 0$. Then M is of the form $z \mapsto z + \beta$ with $\beta \in \mathbb{R}$: It is a translation parallel to the x-axis.

(i)′ $a + d \in \{-2, 2\}$, and $c \neq 0$. M is then parabolic, and the only fixed point $-d/c$ lies in $\mathbb{R}$.

(ii) $|a + d| > 2$ and $c = 0$. M is hyperbolic with fixed points $b/(d - a)$ and ∞ (here necessarily $a \neq d$). M has the form $z \mapsto az/d + b/d$, where $a/d = a^2 > 0$. Invariant lines are all lines through the finite fixed point.

(ii)′ $|a + d| > 2$ and $c \neq 0$. M is hyperbolic, the fixed points

$$\xi_{1,2} = \frac{(a - d) \pm \sqrt{(a + d)^2 - 4}}{2c}$$

lie in $\mathbb{R}$.

(iii) $|a + d| < 2$ and $c = 0$. This cannot occur, because there are no real numbers a, d with $ad = 1$ and $|a + d| < 2$.

(iii)′ $|a + d| < 2$ and $c \neq 0$. M is elliptic with fixed points

$$\xi_{1,2} = \frac{a - d \pm i\sqrt{4 - (a + d)^2}}{2c}.$$

Here, $\xi_1 = \overline{\xi_2}$. $\mathbb{R}$ is an invariant line for M.

(iv) M is not loxodromic.

Proof This follows immediately from the previous results on parabolic, hyperbolic and elliptic transformations.   □

You can imagine these transformations as follows (the black point is transformed into the other gray points):

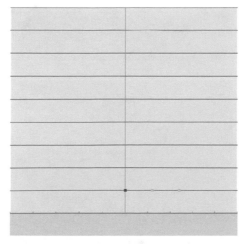

A typical parabolic transformation (fixed point $= \infty$), which leaves $\mathbb{H}$ invariant

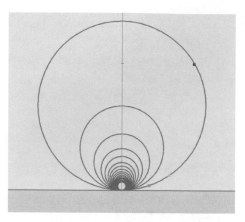

A typical parabolic transformation (fixed point finite), which leaves $\mathbb{H}$ invariant

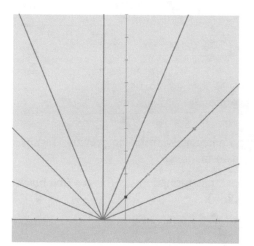

A typical hyperbolic transformation with only one finite fixed point, which leaves $\mathbb{H}$ invariant

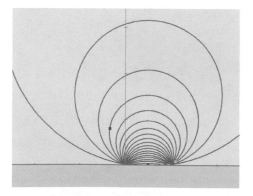

A typical hyperbolic transformation with two finite fixed points, which leaves $\mathbb{H}$ invariant

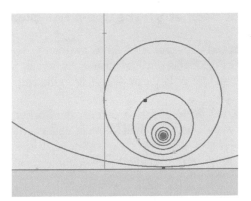

A typical elliptical transformation with finite fixed point, which leaves $\mathbb{H}$ invariant

6.4   Digression: Hyperbolic Geometry

Geometry has been practiced for millennia. Already the Egyptians could solve many practical geometric problems that arose in the construction of pyramids or from the need to remeasure the fields after the floods of the Nile. But for the history of science it is of particular importance that geometry is the first area of mathematics for which an axiomatization was developed. This was carried out in Euclid's "Elements", a work that summarized the knowledge of the time and was created around 300 BC. For many centuries—up to modern times—the "Elements" were used for mathematics teaching. The axiomatic approach was considered exemplary.

Euclid begins with definitions, such as "A point is something that has no parts". Then follow the actual axioms, for example: "Through any two different points there is exactly one line". There is also the so-called *Parallel Axiom*: "For any line G and any point P that does not lie on G, there is exactly one parallel to G through P ". (Two lines are *parallel* if they do not intersect.)

These axioms are something like a concentrate of life experience, and that is why it is hardly surprising that the consequences, that is, the results of geometry, can be applied very well to the solution of concrete problems.

It has long been asked whether Euklid's axiom system is really minimal or whether certain axioms can be derived from others and therefore omitted. This is especially true for the parallel axiom. It was a sensation for the experts when it turned out in the 19th century that there are geometries in which the parallel axiom does not apply: Such geometries are called *non-Euclidean geometries*. The mathematicians Bolyai (1802–1860) and Lobachevsky (1792–1856) are considered the discoverers of the first examples, but it was later found that Gauss (1777–1855) had very similar ideas[3].

Here we want to study a special example of a non-Euclidean geometry, the *hyperbolic geometry*.

---

[3] However, he never published them because he could not imagine that the mathematicians of his time would accept such revolutionary new developments.

6.4.1 Hyperbolic Geometry I: The Upper Half-plane $\mathbb{H}$

One can develop hyperbolic geometry quite abstractly, but if one believes in Euclidean geometry of the plane, one can make a "concrete" image: It can be visualized—according to an idea of Poincaré—in $\mathbb{H}$ or in the unit sphere $\mathbb{U}$.

Basic definitions At the beginning is a definition of what one understands by "point", "line", "between" etc. And then you can check which geometric axioms are fulfilled.

Points. In the model we will consider here, we understand points to be the elements of $\mathbb{H}$.

Lines. A line is either a half-line perpendicular to $\mathbb{R}$ (i.e., a set of the form $\{z \in \mathbb{H} \mid \operatorname{Re} z = c\}$ for some $c \in \mathbb{R}$) or a half-circle with center at $\mathbb{R}$ (i.e., a set of the form $\{z \in \mathbb{H} \mid |z - c| = r\}$ with $c \in \mathbb{R}, r > 0$).

Below one sees some lines. (In order to make the sketch more attractive, the intervening spaces have been colored.)

Some lines in $\mathbb{H}$

Some further definitions are then obvious, for example:

- Two lines *intersect* if they have at least one point in common.
- Two lines are *parallel* if they do not intersect.
- A *line segment* is the piece between two points on a line.
- A *triangle* is a set that is limited by line segments between three points that do not lie on a line. (You can also allow triangles in which one corner is ∞. They look like this: there are two lines perpendicular to the x axis, and at the bottom a piece of a semicircle with center on the x axis is attached.)
- The *angle* between two lines that intersect in z is the angle between the tangents to these lines at z.

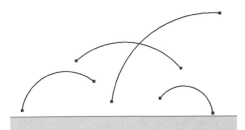

Some line segments in $\mathbb{H}$

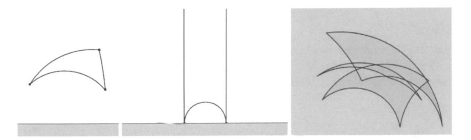

Some triangles in $\mathbb{H}$

Then many results can be easily derived. Some of them we know from Euclidean geometry, some of them have no analogue there. For example:

- Through each two different points passes exactly one line.
- Two not identical lines have at most one point of intersection.
- Is z a point which does not lie on a line G, then through z go infinitely many parallels to G.

Lengths Now we want to measure *distances*. We proceed in *two steps*. First, we define the *length of curves*.

In general, we consider a weight function U on an open subset $\mathbb{R}^n$ of $\rho : U \to \,]0, \infty[$. The number $\rho(x)$ measures how strenuous it is to move forward in the area x. And if $\gamma : [0, 1] \to U$ is a curve[4], then its (weighted by ρ) *length* is defined as $L(\gamma) := \int_0^1 \|\gamma'(t)\| \rho(\gamma(t)) dt$. ($\|\gamma'(t)\|$ stands for the length of the velocity vector at t.) If, for example, ρ is constantly equal to one, then the usual curve length results.

Then calculate, for two given points P and Q in U, the infimum of the curve lengths over all curves from P to Q . This is the *distance* from P to Q. Sometimes there is exactly one curve that realizes the smallest distance. It is then called a *geodesic*.

---

[4] All occurring functions should be piecewise smooth.

Back to $\mathbb{H}$. The relevant weight function there is $z \mapsto 1/\operatorname{Im} z$. This means intuitively that it is particularly strenuous to move forward near the x axis. Therefore, it is plausible that when going from one point in $\mathbb{H}$ to another with the same imaginary part, one should not go parallel to the x axis in order to get from one to the other as favorably as possible. It is probably better to make a "detour up".

Here is a simple *example* to start with. How far is it, for example, from $z = x + iy_1$ to $w = x + iy_2$ if z, w are points lying on top of each other (i. e., $y_1 < y_2$)? We consider the connecting curve, i. e., $\gamma : [0, 1] \to \mathbb{H}$, defined by $t \mapsto z + t(w - z)$. This leads to

$$
L(\gamma) = \int_0^1 \frac{|w - z|}{\operatorname{Im}(z + t(w - z))} \, dt = \int_0^1 \frac{y_2 - y_1}{y_1 + t(y_2 - y_1)} \, dt = \log(y_2/y_1),
$$

and smaller values are obviously not possible.

Now let $M = M_{a,b,c,d}$ be a Möbius transformation that leaves $\mathbb{H}$ invariant: The a, b, c, d are therefore real, and $ac - bd = 1$. M leaves distances invariant:

Proposition 6.4.1
M preserves curve lengths: If γ is a curve in $\mathbb{H}$, then γ and $M \circ \gamma$ have the same length.
It follows that M is an isometry.

Proof The result can be derived from the transformation theorem for integrals. We want to argue here infinitesimally—as was previously common in proofs.

Let $z = x + iy \in \mathbb{H}$ be arbitrary and $\Delta z \in \mathbb{C}$ "very" small. The path from z to $z + \Delta z$ was then weighted with $|\Delta z|/y$.

Now we look at the image of this path under M, it is (in good approximation) the path from $M(z)$ to $M(z) + M'(z)\Delta z$. The length is therefore

$$
|M'(z)| \, |\Delta z| / \operatorname{Im} M(z).
$$

We claim that this value agrees with $|\Delta z|/y$, that is,

$$
|M'(z)| / \operatorname{Im} M(z) = 1/y
$$

holds. Indeed,

$$
|M'(z)| = \left| \frac{1}{(cz + d)^2} \right| = \frac{1}{|cx + d|^2}
$$

as well as

$$
\begin{aligned}
\operatorname{Im} M(z) &= \operatorname{Im} \frac{az + b}{cz + d} \\
&= \operatorname{Im} \frac{(az + b)(c\bar{z} + d)}{|cz + d|^2} \\
&= \operatorname{Im} \frac{(ax + b + iay)(cx - icy + d)}{|cz + d|^2} \\
&= \frac{aycx - aycx + y(ad - bc)}{|cz + d|^2} \\
&= \frac{y}{|cz + d|^2}.
\end{aligned}
$$

Here it was important that a, b, c, d are real.

That M is then an isometry is clear. □

This opens up a way to easily calculate distances. If z, w are given arbitrarily, then choose an ($\mathbb{H}$ invariant) Möbius transformation M so that Mz, Mw have the same real part. The distance is then, as shown a few lines ago, easy to calculate.

Surfaces We weighted lengths with $1/\operatorname{Im} z$ at z. Therefore, it is plausible to weight surface measurements locally with $1/(\operatorname{Im} z)^2$: If G is a subset of $\mathbb{H}$, we define the *surface of G* as

$$
F(G) := \iint_G \frac{1}{(\operatorname{Im} z)^2}\, dx\, dy = \iint_G \frac{1}{y^2}\, dx\, dy.
$$

(One can imagine that $\mathbb{H}$ is a landscape in which we want to purchase a property G. The square meter price varies greatly, it depends inversely quadratically on the distance to the x-axis. $F(G)$ can then be interpreted as the property value.)

Examples:

1) G is bounded by the three Euclidean lines $\{\operatorname{Re} z = a\}$, $\{\operatorname{Re} z = b\}$ and $\{\operatorname{Im} z = c\}$, where $a, b, c \in \mathbb{R}$, $a < b$ and $c > 0$. Then

$$
F(G) = \int_a^b \int_c^\infty \frac{1}{y^2}\, dy\, dx = \int_a^b \frac{1}{c}\, dy = c(b - a).
$$

2) This time we consider the area G, which is bounded by the Euclidean lines $\{\operatorname{Re} z = a\}$, $\{\operatorname{Re} z = b\}$ and the semicircular arc $\{|z| = 1\}$. We assume that $-1 < a < b < 1$. Then

$$
F(G) = \int_a^b \int_{\sqrt{1-x^2}}^\infty \frac{1}{y^2}\, dy\, dx = \int_a^b \frac{1}{\sqrt{1 - x^2}}\, dx = \arcsin(b) - \arcsin(a).
$$

If we write a as $\cos \phi$ and b as $\cos \psi$ and then make the substitution $x = \cos t$, the integral $\int_\phi^\psi t\, dt = \phi - \psi$ results.

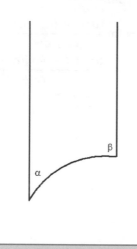

A special triangle in $\mathbb{H}$

Proposition 6.4.2
Möbius transformations that leave $\mathbb{H}$ invariant are area-preserving.

Proof Let M be such a Möbius transformation. An infinitesimal piece of area ΔG at $z = x + iy$ is, however (with an Euclidean interpretationë), enlarged or reduced by $|M'(z)|^2$. From the proof of the previous proposition we conclude that this number has the value $1/|cz + d|^4$. The imaginary part of $M(z)$ is equal to $y/|cz + d|^2$, and therefore $|\Delta G|/y^2$ (up to an infinitesimal error) is equal to $|M(\Delta G)/(\operatorname{Im} M(z))^2$. The mapped area is therefore counted just like the original.   □

This provides the important results needed to prove the following surprising theorem:

Proposition 6.4.3
Let D be a hyperbolic triangle with interior angles α, β, γ. Then

$$F(D) = \pi - (\alpha + \beta + \gamma).$$

In particular, the angle sum in the triangle is always less than π.

Proof The proof idea is to first prove the statement for a simple example: for a triangle like the one in Example 2 (one corner in ∞).

That is a triangle with angles $0, \pi - \phi$ and ψ. There the area is really equal to

$$\phi - \psi = \pi - \big((\pi - \phi) + \psi + 0\big).$$

Thus the statement is true in this case. Here you can see this triangle, where we have set $\alpha = \pi - \phi$ and $\beta = \psi$. These are the proper angles, all you have to do is think that the left lower (or right lower) point of the triangle is equal to $a = e^{i\phi}$ (or equal to $b = e^{i\psi}$).

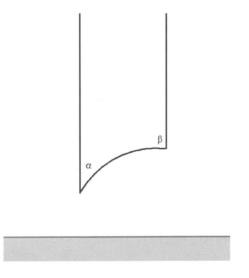

A special triangle in $\mathbb{H}$

Then certain triangles are considered, in which all corners are finite. In the following example we consider the triangle with the corners in a, b, c. The statement is true for the triangle with corners in ∞, a, b and in the triangle with corners in ∞, c, b. And from that it follows by calculating differences that the assertion is also true for the triangle with the corners in a, b, c.

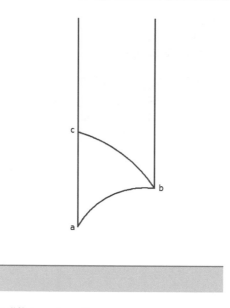

Calculate triangle area by difference formation

Finally, the above proposition is used to solve the general case. It should be noted that Möbius transformations are angle-preserving holomorphic maps.   □

6.4.2   Hyperbolic Geometry II: The Unit Circle $\mathbb{U}$

We have already shown in Sect. 6.3 that the Cayley transformation $\mathcal{C}$ maps the upper half-plane $\mathbb{H}$ to the interior of the unit circle $\mathbb{U}$. Therefore, by conjugation, everything that has been developed for $\mathbb{H}$ can be transferred to $\mathbb{U}$. In particular, this applies to hyperbolic geometry. In the $\mathbb{U}$-variant:

- The points are the elements of $\mathbb{U}$.
- The lines are the circular arcs that intersect orthogonally at the boundary together with the lines through the center.

Here you can see some lines, segments and triangles

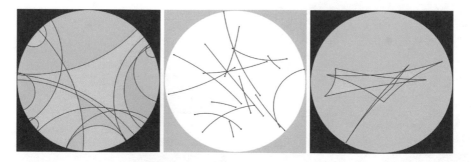

Some lines, segments, and triangles in $\mathbb{U}$

After an appropriate definition of distance, angles, and areas, one can determine the area of a triangle simply by $\pi - (\alpha + \beta + \gamma)$ and prove that Möbius transformations that leave $\mathbb{U}$ invariant preserve lengths and areas. This will not be further elaborated here.

6.5 The Modular Group

The group $SL(2, \mathbb{Z})$ is also called the *modular group*. Depending on whether one works with matrices or Möbius transformations, one deals with matrices with integer entries or with transformations $M_{\alpha, \beta, \gamma, \delta}$ with integer $\alpha, \beta, \gamma, \delta$[5]. Since the beginning of the 19th century, it has played an important role in number theory. It leaves $\mathbb{H}$ invariant, we want to discuss some properties.

It became important when studying *elliptic functions*. These are meromorphic functions f, for which there is a lattice generated by a basis $\mathbf{a}_0, \mathbf{b}_0$, so that f is periodic on this lattice:

$$f(z + n\mathbf{a}_0 + m\mathbf{b}_0) = f(z)$$

for all $n, m \in \mathbb{Z}$.

Two lattices generated by $\mathbf{a}_0, \mathbf{b}_0$ and $\mathbf{a}_0', \mathbf{b}_0'$ are called *equivalent* if they arise from each other by a rotation, that is, if there is an $a \in \mathbb{C}$ so that

$$\mathbf{a}_0'\mathbb{Z} + \mathbf{b}_0'\mathbb{Z} = a(\mathbf{a}_0\mathbb{Z} + \mathbf{b}_0\mathbb{Z}).$$

Obviously, it is enough to concentrate on lattices of the form $\mathbb{Z} + \mathbf{w}\mathbb{Z}$ with $\mathbf{w} \in \mathbb{H}$, and the question arises: When are such lattices $\mathbb{Z} + \mathbf{w}\mathbb{Z}$ and $\mathbb{Z} + \mathbf{w}'\mathbb{Z}$ equivalent? For these investigations you need the modular group:

Proposition 6.5.1
The lattices generated by $\mathbb{Z} + \mathbf{w}\mathbb{Z}$ and $\mathbb{Z} + \mathbf{w}'\mathbb{Z}$ (with $\mathbf{w}, \mathbf{w}' \in \mathbb{H}$) are equivalent exactly if there is a Möbius transformation $M_{\alpha, \beta, \gamma, \delta}$ with $\alpha, \beta, \gamma, \delta \in \mathbb{Z}$ and $\alpha\delta - \beta\gamma = 1$ so that $\mathbf{w}' = M\mathbf{w}$.

Proof First, we assume that $\mathbb{Z} + \mathbf{w}\mathbb{Z} = a(\mathbb{Z} + \mathbf{w}'\mathbb{Z})$ for a $a \neq 0$. This implies $(w', 1)^\top = aM(w, 1)^\top$ for an integer matrix M with entries $\alpha, \beta, \gamma, \delta$. Similarly, $a(1, w)^\top = N(1, w')^\top$ must hold for a matrix N with integer entries. Then one has $(w', 1)^\top = MN(w', 1)^\top$. Since $1, w'$ are linearly independent over $\mathbb{R}$, it follows that $MN = \text{Id}$, and therefore the determinant of M is in $\{-1, 1\}$.

It is even equal to 1. If you divide the two equations of the matrix equation $(w', 1)^\top = aM(w, 1)^\top$ by each other, you get $w' = M_{\alpha, \beta, \gamma, \delta}w$, and since w and w'

---

[5] In the second case, strictly speaking, the factor group $PSL(2, \mathbb{Z}) = SL(2, \mathbb{Z})/\{\pm\text{Id}\}$ is relevant.

are in $\mathbb{H}$, you can conclude that $\det M > 0$. To prove this, we write $w = x + iy$ and calculate as in the proof of proposition 6.4.1:

$$\operatorname{Im} w' = \operatorname{Im} M_{\alpha,\beta,\gamma,\delta} w$$
$$= \frac{y(\alpha\delta - \beta\gamma)}{|\gamma w + \delta|^2}.$$

Since the left side and y are positive, $\alpha\delta - \beta\gamma > 0$ must also hold.

And now let $w' = M_{\alpha,\beta,\gamma,\delta} w$ for an $M_{\alpha,\beta,\gamma,\delta}$ with $\alpha, \beta, \gamma, \delta \in \mathbb{Z}$ and $\alpha\delta - \beta\gamma = 1$. Set $a := 1/(\gamma w + \delta)$. For the matrix M with entries $\alpha, \beta, \gamma, \delta$, it then follows that $(w', 1)^\top = aM(w, 1)^\top$, and this immediately implies $\mathbb{Z} + w'\mathbb{Z} = a(\mathbb{Z} + w\mathbb{Z})$.   □

Another motivation arose from the study of *quadratic forms* with integer coefficients. For example: which integers can be written in the form $5x^2 - 12xy + y^2$ with $x, y \in \mathbb{Z}$? In general, let B be a 2×2-matrix with integer entries, and we investigate the $\langle \mathbf{x}, B\mathbf{x} \rangle$ for $\mathbf{x} \in \mathbb{Z}^2$. If $A \in SL(2, \mathbb{Z})$, then $A\mathbf{x}$ exhausts $\mathbb{Z}^2$, i.e. instead of $\langle \mathbf{x}, B\mathbf{x} \rangle$ one can also investigate $\langle A\mathbf{x}, BA\mathbf{x} \rangle = \langle \mathbf{x}, A^\top BA\mathbf{x} \rangle$. Or expressed differently: B can be replaced by $A^\top BA$, and thereby the problem may be simplified.

For example, let B be the identity matrix and $A = \left(\begin{smallmatrix} 2 & 1 \\ 1 & 1 \end{smallmatrix}\right)$. Then $A^\top BA = \left(\begin{smallmatrix} 5 & 3 \\ 3 & 2 \end{smallmatrix}\right)$. So the problem "Find for given c all integer solutions of $5x^2 + 6xy + 2y^2 = c$" is reduced to "Find for given c all integer solutions of $x^2 + y^2 = c$ ". For $d = 157$, for example, the solution $(11, 6)^\top$ is found, and $A(x, y)^\top = (11, 6)^\top$ yields $(x, y) = (5, 3)$. *This* tuple solves $5x^2 + 6xy + 2y^2 = 157$.

Let S, T be the following elements of the $SL(2, \mathbb{Z})$:

$$S = \begin{pmatrix} 1 & 1 \\ 0 & 1 \end{pmatrix} \quad \text{and} \quad T = \begin{pmatrix} 0 & 1 \\ -1 & 0 \end{pmatrix}.$$

They correspond to the Möbius transformations $z \mapsto z + 1$ or . We claim:

Proposition 6.5.2
S, T generate the group $SL(2, \mathbb{Z})$.

Proof We first notice that for all $M = \left(\begin{smallmatrix} a & b \\ c & d \end{smallmatrix}\right)$ one has

$$MS = \begin{pmatrix} a & a+b \\ c & c+d \end{pmatrix}, \quad MT = \begin{pmatrix} -b & a \\ -d & c \end{pmatrix}, \quad MS^{-1} = \begin{pmatrix} a & b-a \\ c & d-c \end{pmatrix}.$$

Now let $M \in SL(2, \mathbb{Z})$ be given arbitrarily. With $\mathbb{M}$ we denote the set of transformations that arise from M by multiplying elements from $\{S, T, S^{-1}, T^{-1}\}$ from the right arbitrarily often. It will suffice to show that $\operatorname{Id} \in \mathbb{M}$, because then M can be written as a product of powers of the elements from $\{S, T, S^{-1}, T^{-1}\}$.

In a first step we define $\tau \in \mathbb{N}_0$ as the minimal $|a'|$ of the $\left(\begin{smallmatrix} a' & b' \\ c' & d' \end{smallmatrix} \right) \in \mathbb{M}$. We claim that $\tau = 0$ holds, i. e. that there is a matrix $\left(\begin{smallmatrix} 0 & b' \\ c' & d' \end{smallmatrix} \right)$ in $\mathbb{M}$. For the proof we choose a $M' = \left(\begin{smallmatrix} a' & b' \\ c' & d' \end{smallmatrix} \right) \in \mathbb{M}$ with $|a'| = \tau$ and assume that $\tau > 0$ holds. (Our aim: A contradiction follows.)

Case 1: $|b'| < |a'|$.

This cannot be, because $M'T$ lies in $\mathbb{M}$, and this matrix would have the element $-b'$ in the upper left, thus an element with absolute value smaller than τ.

Case 2: $|b'| \geq |a'| > 0$.

This also leads to a contradiction. If a', b' have different (resp. the same) sign, then apply S (resp. S^{-1}) to M' possibly several times from the right. This would create an element from $\mathbb{M}$ for which case 1 applies.

(For example, if $M' = \left(\begin{smallmatrix} -2 & 5 \\ * & * \end{smallmatrix} \right)$, then $M'S^2 = \left(\begin{smallmatrix} -2 & 1 \\ * & * \end{smallmatrix} \right)$.)

Consequently there is an $M' = \left(\begin{smallmatrix} 0 & b' \\ c' & d' \end{smallmatrix} \right)$ in $\mathbb{M}$. Necessarily one has $-b'c' = 1$.

Case 1: $b' = 1$, $c' = -1$. Again, by repeatedly multiplying M' by S or S^{-1}, one can achieve that d' becomes zero, and that means $T \in \mathbb{M}$ and therefore $\text{Id} = T^4 \in \mathbb{M}$.

Case 2: $b' = -1$, $c' = 1$. This time, $-T \in \mathbb{M}$ follows in a similar way, and because of $T^2 = -\text{Id}$, $T \in \mathbb{M}$ and therefore $\text{Id} \in \mathbb{M}$ are also true.   □

To conclude this section, we want to identify one more fundamental domain of the modular group. We set

$$\mathbb{F} := \{ z \in \mathbb{H} \mid -0{,}5 \leq \text{Re}\, z \leq 0{,}5, \ |z| \geq 1 \}$$

and claim:

Proposition 6.5.3

$\mathbb{F}$ is a fundamental domain for the modular group operating on $\mathbb{H}$.

Proof We have to show:

a) For each $z \in \mathbb{H}$ there is a matrix $\left(\begin{smallmatrix} a' & b' \\ c' & d' \end{smallmatrix} \right) \in SL(2, \mathbb{Z})$ with $M_{a',b',c',d'}\, z \in \mathbb{F}$.

b) The by the $T \in SL(2, \mathbb{Z})$ shifted copies of $\mathbb{F}$ do not intersect in the interior.

Let $z \in \mathbb{H}$ be given. We determine $c, d \in \mathbb{Z}$ so that $|cz + d|$ is strictly positive and minimal. Then c, d are coprime, and we can choose $a, b \in \mathbb{Z}$ so that $ad - bc = 1$. Set $M = \left(\begin{smallmatrix} a & b \\ c & d \end{smallmatrix} \right)$. This matrix belongs to $SL(2, \mathbb{Z})$, and for each $m \in \mathbb{Z}$ one has $S^m M = \left(\begin{smallmatrix} a+mc & b+md \\ c & d \end{smallmatrix} \right)$. First, due to our choice of c, d

$$|M_{a+mc, b+md, c, d}\, z| = \left| \frac{(a + mc)\, z + (b + md)}{cz + d} \right| \geq 1.$$

Second, $M_{a+mc,b+md,c,d} z = m + M_{a,b,c,d} z$, and therefore one can by suitable choice of m achieve that the real part of $M_{a+mc,b+md,c,d} z$ lies between $-1/2$ and $1/2$. Then $|M_{a+mc,b+md,c,d} z| \geq 1$ and

$$-0.5 \leq \operatorname{Re} M_{a+mc,b+md,c,d} z \leq 0.5,$$

i.e. $M_{a+mc,b+md,c,d} z \in \mathbb{F}$.

In the second part of the proof we have to show that non-trivially transformed images of $\mathbb{F}$ do not intersect in the interior. We have to show: If $M_{a,b,c,d}$ is a Möbius transformation with $a, b, c, d \in \mathbb{Z}$ and $ad - bc = 1$, so that a $z \in \mathbb{F}^0$ exists for which also Mz belongs to $\mathbb{F}^0$, then $M = \mathrm{Id}$.

Let M and $z = x + iy$ be given. We notice that the lower corners of $\mathbb{F}$ have the smallest imaginary part, so that $t \geq \sqrt{3}/2$ for all $s + it \in \mathbb{F}$. In particular,

$$\operatorname{Im} Mz = \frac{y}{|cz + d|^2} \geq \frac{\sqrt{3}}{2}.$$

holds On the other hand, $|cz + d|^2 \geq |\operatorname{Im}(cz + d)|^2 = c^2 y^2$, and it follows

$$\frac{\sqrt{3}}{2} \leq \frac{y}{|cz + d|^2} \leq \frac{y}{c^2 y^2} = \frac{1}{c^2 y} \leq \frac{1}{c^2} \frac{2}{\sqrt{3}}.$$

This proves $c^2 \leq 4/3$, i.e., $c \in \{0, 1, -1\}$.

First, let $c = 0$. Then, because of $ad = 1$, it is a Möbius transformation of the form $w \mapsto w + m$ for an $m \in \mathbb{Z}$. Only for $m = 0$, that is $M = \mathrm{Id}$, no contradiction arises.

There remain the cases $c = \pm 1$, and w.l.o.g we may assume $c = 1$. (Otherwise, we multiply all entries of with -1.) M is therefore of the form $Mw = (aw + b)/(w + d)$. How large can d be? For $\alpha, \beta \geq 0$, $\alpha^2 + \beta^2 \geq 2\alpha\beta$ always holds, which follows from $(\alpha - \beta)^2 \geq 0$. So $|z + d|^2 = |x + d|^2 + |y|^2 \geq 2|x + d|y$ and thus

$$\frac{\sqrt{3}}{2} \leq \frac{y}{|c + d|^2} \leq \frac{y}{2|x + d|y} = \frac{1}{2|x + d|}.$$

This implies $\sqrt{3}|x + d| \leq 1$, and because of $|x| \leq 0.5$ only the possibilities $d = 0$ and $d = \pm 1$ remain.

First, we consider the case $d = 0$. Then it is necessary that $b = -1$, so it is the transformation $a - 1/w$, and it is certainly not possible that z and $a - 1/z$ belong to the interior of $\mathbb{F}$. (This is easily seen by considering the cases $a = 0$ or $a \neq 0$.)

We still have to treat the case $d = \pm 1$. One has $|z + d|^2 = (x \pm 1)^2 + y^2$, so $|z + d|^2 \geq 0.25 + y^2$, since $|x| \leq 0.5$. Consequently,

$$\frac{\sqrt{3}}{2} \leq \operatorname{Im} Mz = \frac{y}{|z + d|^2} \leq \frac{y}{y^2 + 0.25}.$$

In this way the function $\phi(y) := y/(y^2 + 0{,}25)$ comes into play, which only is of importance for us in the area $y \geq \sqrt{3}/2$. There it is strictly monotonously decreasing (the derivative is negative), and at $\sqrt{3}/2$ it has the value $\sqrt{3}/2$. The only possibility for y is therefore $y = \sqrt{3}/2$, but then z does not lie inside $\mathbb{F}$. □

How $\mathbb{H}$ is filled by the $T(\mathbb{F})$ when T runs through the module group can be seen in the following picture:

A fundamental domain of the modular group ($\mathbb{F}$ is above in the middle)

Of course, you can also make the fundamental domain a little more interesting by using more colors or by inserting pictures:

A tiling of $\mathbb{H}$ induced by the modular group

Also, you can conjugate everything by the Cayley transformation into the $\mathbb{U}$ world (cf. Sect. 6.4.2). Then, using the conjugated fundamental domain $\mathbb{F}$, a tiling of $\mathbb{U}$ results, which is shown below in two variants:

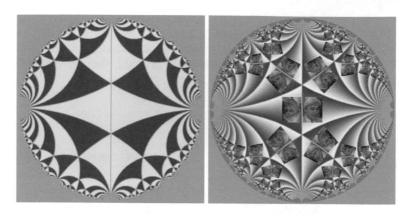

A tiling of $\mathbb{U}$ induced by the modular group

6.6   Groups with Two Generators

We briefly interrupt the study of Möbius transformations to remind ourselves of some terms and facts from group theory.

Particularly easy are the *cyclic groups* G, where there is a g so that G agrees with $\{g^n \mid n \in \mathbb{Z}\}$. They are always commutative, and they can be finite or countable. They are also said to be *generated* by a single element. The countable case, for example $\mathbb{Z}$, plays a special role, because all cyclic groups arise from such groups as quotients.

The modular group treated in the previous section is, as we have seen, a group with *two generators*. In general, $g_1, \ldots, g_k$ are called *generators* of a group G if G is the smallest subgroup containing these elements. It is not difficult to see that each element of G then can be written as $h_1 \circ \ldots \circ h_r$, where $r \in \mathbb{N}$ and all h_i are in $\{g_1, \ldots, g_k, g_1^{-1}, \ldots, g_k^{-1}\}$. If such $g_1, \ldots, g_k$ exist, the group is called *finitely generated*.

Groups with two generators are obviously the next simplest case after cyclic groups. We will see in the following sections examples of groups of Möbius transformations that already give rise to extremely difficult problems. Just as in the case of cyclic groups, where $\mathbb{Z}$ in a certain sense is the *most general* such group (any other is a quotient), there is also a "particularly general" one in the case of groups with k generators.

We want to describe their properties for the special case $k = 2$. First to the *definition*. Under, the *free group with 2 generators*, we understand

- the set of finite words formed from the symbols a, A, b, B, in which never aA, Aa, bB, or Bb occurs,
- together with the symbol e.

Here you can see some elements: $abbaBBBBB, e, Ba, bbbbbbbbbAAbABa, \ldots$ *Not to F_2 belong for example $aabBaaA$ or eA.*

The group operation can be easily memorized. "Secretly" is $A = a^{-1}$ and $B = b^{-1}$, and e is the neutral element. Formally, one defines:

- $e \circ x = x \circ e = x$ for any x.
- If x, y are any elements of F_2, one writes the associated finite words next to each other to define $x \circ y$. And then one simplifies, all letter combinations aA, Aa, bB or Bb are cancelled, this action may have to be repeated several times. If nothing is left at the end, e should be the result.

Then one can show that the group multiplication is first well-defined and secondly defines a group structure on F_2 . So, for example, $a \circ A = e$, $aabA$ is inverse to $aBAA$, etc. And F_2 is generated by the two elements a, b.

Every quotient of F_2 by a normal subgroup is also generated by two elements (perhaps even by one), one need only consider the elements $[a], [b]$ in the quotient. Remarkably, the converse also holds:

Proposition 6.6.1
Let $(G, \circ)$ be a group generated by the elements g, h. Then G is isomorphic to a quotient group of F_2.

Proof Define $\phi : F_2 \to G$ by $e \mapsto e_G$, $a \mapsto g$, $b \mapsto h$ and for the other elements of F_2 such that ϕ is a group homomorphism:

$$A \mapsto g^{-1}, \ B \mapsto h^{-1}, \ aabABa \mapsto gghg^{-1}h^{-1}g,$$

etc. ϕ is then a well-defined group homomorphism, which is onto by assumption (g, h generate G). By the homomorphism theorem, G is isomorphic to the quotient of F_2 by the kernel of ϕ. $\square$

In this way, one can construct groups with tailor-made properties, for example one generated by g, h where in addition $g^7 = e$ and $ghg = h^2$ hold. There one considers in F_2 the normal subgroup generated by a^7 and $abaBB$ and passes to F_2/N.

It should also be clear how to generalize the construction to more than two (even to an arbitrary number) of generating elements. This leads to the free groups with k generators, where k can be any cardinal number.

So far, we have only learned about cyclic groups and the modular group as subgroups of Möbius transformations. The latter is *not* free, because there are non-trivial relations between the generators, such as (with the notation introduced in the corresponding chapter) $T^4 = \mathrm{Id}$ or $(ST)^3 = \mathrm{Id}$.

6.7   Schottky Groups

The modular group was generated by the two maps ($z \mapsto z + 1$ and $z \mapsto -1/z$). We will now study other groups of Möbius transformations with two generators, which will be relatively easy to visualize. They are named after *Friedrich Schottky* (1851–1935), who first described them in 1879.

In preparation, we fix two circles[6] K_1, K_2 in the plane, for example the ones pictured here (K_1 on the left, K_2 on the right):

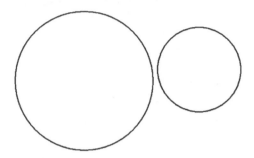

Two circles in the plane

The left one has the center m_1 and the radius r_1, the right one has the center m_2 and the radius r_2. We only require that the circles do not intersect, that is, $|z_1 - z_2| > r_1 + r_2$ holds.

We are looking for a Möbius transformation M that does the following. First, it maps the boundary of K_1 to the boundary of K_2. There are then two possibilities: The interior of K_1 can be mapped to the interior or the exterior of K_2. We vote for the second possibility, which means that the singularity of M is in K_1. An example for such a M is quickly found[7], we set

$$Mz := \frac{m_2 z + (r_1 r_2 - m_1 m_2)}{z - m_1} = \frac{r_1 r_2}{z - m_1} + m_2.$$

Really, the singularity is at $m_1 \in K_1$, and from $|z - m_1| = r_1$ it follows that

$$|Mz - m_2| = \left| \left(\frac{r_1 r_2}{z - m_1} + m_2 \right) - m_2 \right|$$

$$= \left| \frac{r_1 r_2}{z - m_1} \right|$$

$$= r_2 \,;$$

the boundary of K_1 is mapped to the boundary of K_2.

---

[6] By "circles" we mean here the whole circular discs, not just the boundary.

[7] Depending on which point from K_1 is mapped to ∞, there is another wish left.

In order to better understand M, we consider two "test circles": one gray and one light green:

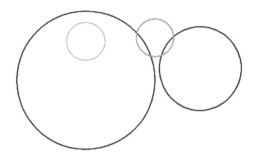

Two test circles: *gray* and *light green*

The gray one lies inside of K_1, so it must be mapped to the outside of K_2. The light green one intersects the outside and the inside of K_1, so the image must also touch the outside and the inside of K_2. This is in fact the case, in the next picture you can see the image under M of the gray (or light green) circle as a black (or dark green) circle:

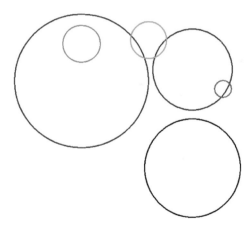

The images under M of the test circles: *black* and *dark green*

And how does M^{-1} work? Obviously, the boundary of K_2 is mapped to the boundary of K_1, and the interior or exterior of K_2 lands in the exterior or interior of K_1. Here you can see the effect of M^{-1} on our test circles:

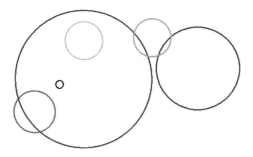

The images under M^{-1} of the test circles: again *black* and *dark green*

A fundamental domain F for the group generated by M is also quickly found: one can choose the complement of the two circles. Here you see F (light green) and next to it F together with $M(F)$ (blue) and $M^{-1}(F)$ (red).

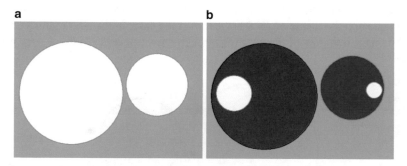

a the fundamental domain F; **b** F and $M(F)$ (*blue*) and $M^{-1}(F)$ (*red*)

Note that the blue and red areas are also the complement of two discs (in the previously white area), respectively.

This can be iterated. The $M^n(F)$ lie for $n \in \mathbb{N}$ all in K_2, the $M^{-n}(F)$ in K_1, and if one lets n run to infinity, the whole complex plane is filled with the exception of two points (the fixed points of M).

So far our preparation. In a *Schottky group* now *two* of the above described constellations are considered simultaneously. More precisely, there are

- four pairwise disjoint circles K_1 and K_2 as well as K_1' and K_2';
- a Möbius transformation M, which maps the boundary of K_1 to the boundary of K_2 and the interior of K_1 to the exterior of K_2;
- a Möbius transformation M', which maps the boundary of K_1' to the boundary of K_2' and the interior of K_1' to the exterior of K_2'.

The associated *Schottky group* $\mathcal{G}$ is then the subgroup of Möbius transformations generated by M, M'.

Because of our preparations, we know how the M, M' acts on our circles:

M maps K_1', K_2' and K_2 to the interior of K_2; the boundary of K_1 is mapped to the boundary of K_2.

M^{-1} maps K_1', K_2' and K_1 to the interior of K_1; the boundary of K_2 is mapped to the boundary of K_1.

M' maps K_1, K_2 and K_2' to the interior of K_2'; the boundary of K_1' is mapped to the boundary of K_2'.

M'^{-1} maps K_1, K_2 and K_1' to the interior of K_1'; the boundary of K_2' is mapped to the boundary of K_1'.

We can visualize this as follows. First we see the four circles, on the left as ordinary circles and on the right slightly decorated graphically:

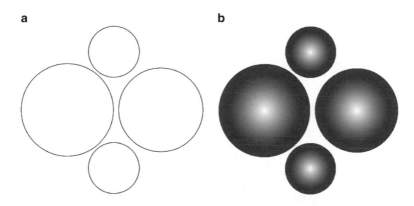

a The starting circles of a Schottky group (in **b** graphically "decorated")

And then the images of the 4 circles under the 4 Möbius transformations M, M^{-1}, M', M'^{-1} (in red):

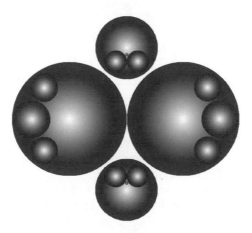

The images of the 4 circles under M, M^{-1}, M', M'^{-1} (*red*)

This can be iterated. If one transforms all previously constructed circles again under M, M^{-1}, M', M'^{-1}, the following picture results:

M, M^{-1}, M', M'^{-1} are applied again (new circles in *yellow*)

And once again:

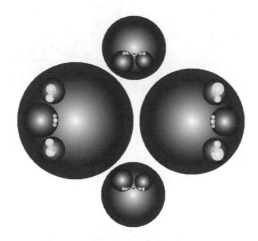

M, M^{-1}, M', M'^{-1} are applied again (new circles in *light green*)

This can also be expressed differently. Each element of the group generated by M, M' is of the form $M_1 \circ \cdots \circ M_k$, where $M_i \in \{M, M^{-1}, M', M'^{-1}\}$ and $k \in \mathbb{N}$. (If M and M^{-1} or M' and M'^{-1} are next to each other, both can of course be omitted.) In the previous pictures, all group elements were applied to all circles with $k \leq 3$, first $k \leq 1$, then $k \leq 2$ and finally $k \leq 3$.

In the last image, however, the circle images are already so tiny that the radii are smaller than the resolution of the image in many cases.

You can of course continue this, to $k = 4, 5, 6$ and so on. The circles become smaller and smaller and for $k \to \infty$ they converge to individual points. We will formulate this more precisely in a moment. But first we notice that it is also easy to find a fundamental domain for the Schottky group considered here, one only has to choose the complement of the four circles:

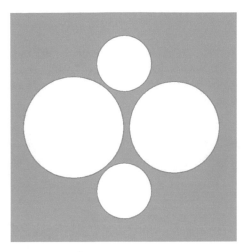

A fundamental domain (*light green*) for the Schottky group considered here

And if you map the fundamental domain under M, M^{-1}, M', M'^{-1}, you get the following picture:

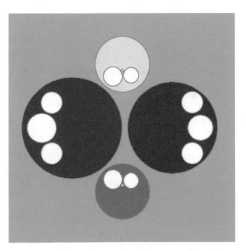

M, M^{-1}, M', M'^{-1} act on this fundamental domain

In the next step (by applying M^2 etc.), the white gaps would be filled by similar shapes, and what finally remains is a quite complicated subset of $\mathbb{C}$, which we will discuss below.

We need some preparations. First, we consider the special Möbius transformation

$$M_u z := \frac{\sqrt{1+u^2}z + u}{uz + \sqrt{1+u^2}}$$

for $u \in \mathbb{R}$. By proposition 6.3.2 $\mathbb{U}$ is mapped onto itself, and since the entries are real, $\overline{Mz} = M\overline{z}$ holds. It follows that a circle in $\mathbb{U}$ with center at $\mathbb{R}$ is mapped to a circle with center also in $\mathbb{R}$. We claim that one can move such circles almost arbitrarily by using suitable u:

Lemma 6.7.1

Let K be a circle with center $m \in \mathbb{R}$ and radius r inside $\mathbb{U}$. Then there is a $u \in \mathbb{R}$ so that $M_u K$ is concentric to the unit circle of $\mathbb{U}$.

Proof Let $s := m - r, t := m + r$, so $-1 < s < t < 1$. These are the points at which the circle intersects the real axis. We want to find a u so that $M_u s + M_u t = 0$. Then, due to the preliminary remark, $M_u K$ is a circle with center 0, so it has the desired properties.

The equation $-M_u s = M_u t$ is equivalent to

$$\frac{s+t}{st+1} = -\frac{2u\sqrt{1+u^2}}{1+2u^2}.$$

The expression on the right goes to ± 1 for $u \to \mp\infty$. So such a u can be found if the number on the left is less in magnitude than 1. This is the case if $(s+t)^2 < (1+st)^2$, that is, if $s^2 + t^2 < 1 + s^2 t^2$. But this is true: It should only be noted that the lines $x \mapsto s^2 + x$ and $x \mapsto 1 + s^2 x$ only intersect at $x = 1$. Thus for t^2 the proper inequality still applies.   □

Lemma 6.7.2

Let D and E be concentric circles, so that D lies inside E. Let r, r' be the radii, so $r < r'$.

Now let M be a Möbius transformation without a singularity in E. The radii of the circles $M(D), M(E)$ should be called s, s'. We claim that $s/s' \le r/r'$.

Proof M can be constructed from translations, rotations and $1/z$. The statement is clear for translations and rotations. It remains to investigate the case $1/z$. After suitable scaling, the center of D and E is equal to $m > 0$, and one has $r' = \eta r$ with an $\eta > 1$, where $m - \eta r > 0$. We map the circles using the map $1/z$. The points $m - r$ and $m + r$ from D are mapped to $1/(m \pm r)$, so the radius of the image sphere is

$$\frac{1}{2}\left(\frac{1}{m+r} - \frac{1}{m-r}\right) = \frac{r}{m^2 - r^2}.$$

Similarly, the radius of the image sphere of E is equal to

$$\frac{1}{2}\left(\frac{1}{m+\eta r} - \frac{1}{m-\eta r}\right) = \frac{\eta r}{m^2 - \eta^2 r^2}.$$

For the quotients of the radii of the images we thus have

$$\frac{s}{s'} = \frac{r}{m^2 - r^2} \cdot \frac{m^2 - \eta^2 r^2}{\eta r} = \frac{1}{\eta}\frac{m^2 - \eta^2 r^2}{m^2 - r^2} < \frac{1}{\eta} = \frac{r}{r'}. \qquad \square$$

Proposition 6.7.3
Let D and E be discs such that D lies inside E. Then there is an $L < 1$ with the following property: If M is a Möbius transformation that has no singularity in E, then the quotient "radius of $M(D)$ divided by radius of $M(E)$" is less than or equal to L.

Proof First, transform E, F by rotation and translation so that E is the unit circle and D is a circle with center on the real axis. Then, using the previous lemma, translate so that the circles are concentric. The resulting transformation is called M_1, and $L < 1$ is the ratio of the radii. Now let M be as in the proposition. We write M as the product of the Möbius transformations MM_1^{-1} and M_1. Then it follows:
Ratio of circle radii $M(D)$ and $M(E)$
$=$ Ratio of circle radii $MM_1^{-1}M_1(D)$ and $MM_1^{-1}M_1(E)$
$\leq$ Ratio of circle radii $M_1(D)$ and $M_1(E)$ (Lemma 6.7.2)
$= L$ (Definition).
Thus the proposition is proven. $\qquad \square$

We want to use the above result to explain the observation that the image circles "get smaller quickly". To do this, we first introduce some terminology. The circles K_1, K_2, K_1' and K_2' should now be called D_A, D_a, D_B, D_b, and M, M^{-1}, M', M'^{-1} are now a, A, b, B. (From here on we follow the terminology proposed in the book by

Mumford et al., which is very suggestive.) This ensures: The non-singular images of the transformations $T \in \{a, A, b, B\}$ are in D_T. So, for example, M is singular on K_1, and the other discs are mapped according to K_2. Therefore, M is now called a, and K_2 is called D_a.

If $c_1, \ldots, c_k \in \{a, A, b, B\}$, then $D_{c_1,\ldots,c_k}$ should denote the circle $c_1 \cdots c_{k-1} D_k$. E.g., $D_{aBaaBA} = aBaaB(D_A)$.

We will only consider those $c_1, \ldots, c_k$ for which a, A or b, B do not occur as neighbours, because we want to stay in the non-singular area.

The $D_{c_1,\ldots,c_k}$ are therefore precisely the circles that we (for not too large k) have drawn in the above images.

How do their radii behave with increasing k? Write $r(D_{c_1,\ldots,c_k})$ for the radius of $D_{c_1,\ldots,c_k}$. We note:

- a maps D_α for $\alpha = a, b, B$ into the interior of D_a;
- A maps D_α for $\alpha = A, b, B$ into the interior of D_A;
- b maps D_α for $\alpha = a, A, b$ into the interior of D_b;
- B maps D_α for $\alpha = a, b, B$ into the interior of D_B.

Consequently, there is an $L < 1$ by which the ratio of the radii (smaller circle through larger circle) can be estimated for these 12 situations. Let R be the maximum of the radii of the D_α. Then we can apply the previous proposition by using a telescope trick. E.g.,

$$
\begin{aligned}
r(D_{aBab}) &= r(aBaD_b) \\
&= \frac{r(aBaD_b)}{r(aBD_a)} \frac{r(aBD_a)}{r(aD_B)} \frac{r(aD_B)}{r(D_a)} r(D_a) \\
&\leq \frac{r(aD_b)}{r(D_a)} \frac{r(BD_a)}{r(D_B)} \frac{r(aD_B)}{r(D_a)} r(D_a) \\
&\leq L^3 R.
\end{aligned}
$$

(The proposition was applied when the aB or a was omitted in the numerator and denominator). And quite analogously one can show that $r(D_{c_1,\ldots,c_k}) \leq L^k R$ for all of the $c_1, \ldots, c_k$.

The radii go to zero "fast", even geometrically fast, as the pictures suggest. We want to investigate what happens for $k \to \infty$.

First, we remind you of the *Cantor intersection theorem*: If (M, d) is a complete metric space, and if $E_1 \supset E_2 \supset \cdots$ are non-empty closed subsets whose diameters tend to zero, then $\bigcap E_i$ consists of exactly one point.

Here we apply this theorem as follows. If we choose sequences $c_1, c_2, \ldots$ in $\{a, A, b, B\}$ so that all $D_{c_1,\ldots,c_k}$ are well-defined, then we denote by E_k the circle corresponding to $D_{c_1,\ldots,c_k}$. By construction, $E_1 \supset E_2 \supset \cdots$: So, for example, $D_{aaBa} = aaB(D_a) \subset aa(D_B) = D_{aaB}$, since $B(D_a) \subset D_B$. The diameters of the E_k tend (geometrically fast) to zero, and therefore there is exactly one point $z_{c_1,c_2,\ldots}$ that lies in all E_k.

These points are called *limit points* of the group $\mathcal{G}$ of Möbius transformations generated by a, b (i.e. M_1, M_2), and the set of all limit points is called the *limit set*. We will call it $\mathbb{L}$. $\mathbb{L}$ has some remarkable properties:

Proposition 6.7.4

For the set $\mathbb{L}$ the following holds:

(i) $\mathbb{L}$ is not empty and closed, and $\mathbb{L}$ is invariant under all $M \in \mathcal{G}$ (i. e. $M(\mathbb{L}) \subset \mathbb{L}$).

(ii) $\mathbb{L}$ contains the fixed points of all nontrivial $M \in \mathcal{G}$, and the set of all these fixed points is dense in $\mathbb{L}$.

(iii) $\mathbb{L}$ is the smallest invariant set: If K is not empty, closed and $M(K) \subset K$ holds for all $M \in \mathcal{G}$, then $\mathbb{L} \subset K$.

(iv) $\mathbb{L}$ is uncountable.

(v)   Earlier we had defined a fundamental domain F (the complement of the circles). $\mathbb{L}$ is the complement of $\bigcup_{M \in \mathcal{G}} M(F)$.

Proof (i) Let C_k be the union over all admissible $D_{c_1, \dots, c_k}$. (Here we interpret the D's as discs.) This is a closed set, and $\mathbb{L} = \bigcap_k C_k$.

Now let $z_{c_1, c_2, \dots} \in \mathbb{L}$ and $M = c_1' \cdots c_r'$ be any element from $\mathcal{G}$. Consider the sequence $c_1', \dots, c_r', c_1, c_2, \dots$. It may have to be "cleaned up" (omit aA, etc.), the result is called $d_1, d_2, \dots$.

Then $M z_{c_1, c_2, \dots} = z_{d_1, d_2, \dots}$.

(ii) Let $M = c_1 \cdots c_k \in \mathcal{G} \setminus \{\mathrm{Id}\}$ be given. Then for each z the limit of $M^n z$ for $n \to \infty$ is equal to $z_{c_1, \dots, c_k, c_1, \dots, c_k, \dots} \in \mathbb{L}$. This limit is the attracting fixed point (except if z is the other fixed point). If you replace M with M^{-1}, it shows that the second fixed point is also in $\mathbb{L}$.

(iii) In part (i) we have already seen that $\mathbb{L}$ is invariant. Now let K be another invariant set, $z \in K$ and $z_{c_1, c_2, \dots} \in \mathbb{L}$. The $M_{c_1, \dots, c_n} z$ converge for $n \to \infty$ to $z_{c_1, c_2, \dots}$, and all $M_{c_1, \dots, c_n} z$ are contained in K. Since K is closed, it follows that $z_{c_1, c_2, \dots} \in K$. This proves $\mathbb{L} \subset K$.

(iv) The elements of $\mathbb{L}$ can still be identified with the admissible sequences $c_1, c_2, \dots$. There are at most as many of them as $\mathbb{R}$ has elements, but also at least as many, because all sequences consisting only of a's and b's are contained.

(v) That is clear. $\square$

Corollary 6.7.5

(i) $\mathcal{G}$ is isomorphic to the free group with two generators.

(ii) $\mathcal{G}$ is a discrete group.

(iii) No $U \in \mathcal{G}$ is elliptic.

Proof (i) Let $c_1, c_2, \ldots, c_k$ and $d_1, d_2, \ldots, d_r$ be two different admissible sequences (so they don't contain a $\alpha\alpha^{-1}$). Then the Möbius transformations $c_1 c_2 \cdots c_k$ and $d_1 d_2 \cdots d_r$ are also different, because one can find points that are mapped to different elements. (For example, $abAB \neq abAb$, because a z that lies outside all circles is mapped under $abAB$ to D_{abAB} and under $abAb$ to D_{abAb}, and D_{abAB} and D_{abAb} are disjoint.) Therefore, the mapping $c_1 \cdots c_k$ (considered as a Möbius transformation) $\mapsto c_1 \cdots c_k$ (considered as an element of F_2) from $\mathcal{G}$ to F_2 is bijective. It is also clear that it is a group morphism.

(ii) This is clear, because there are many points that are not limit points. (For example, points that lie outside the four discs are mapped under all non-trivial transformations of the group to one of the four discs.)

(iii) Let $M = c_1 \cdots c_k \in \mathcal{G}$ be arbitrary. Then for each z the $M^n z$ converge for $n \to \infty$ to $z_{c_1,\ldots,c_k,c_1,\ldots,c_k,\ldots}$. But for elliptic transformations, $(M^n z)$ is only convergent for the fixed points. □

How can one imagine the limit set $\mathbb{L}$? Exactly this is not possible, because a discrete cloud of points does not leave any traces even in the finest resolution of a picture. But *approximately* this can be done, there are even three possibilities:

Representation 1: Determine the $D_{c_1,\ldots,c_k}$ for all admissible $c_1, \ldots, c_k$ and a "sufficiently large" k. Then draw the union of these annuli. $\mathbb{L}$ is a subset, and if k is large enough, it is a good approximation.

Representation 2: Calculate the fixed points for "many" $M \in \mathcal{G}$ and enter them as points in an image. Due to part (ii) of the previous proposition, this is an approximation of $\mathbb{L}$.

Representation 3: Find any fixed point z_0 for any $M \in \mathcal{G} \setminus \{\mathrm{Id}\}$. It is located in $\mathbb{L}$ and is sketched. Then define $z_2, z_3, \ldots$ recursively by $z_{n+1} = M z_n$, where M is chosen randomly in $\mathcal{G}$ each time. Sketch a sufficiently large number of these z_n. All z_n live in $\mathbb{L}$, and if n is large enough, $\mathbb{L}$ should be well approximated.

The pictures generated in this way are hardly distinguishable from each other. Here one sees an approximate image of $\mathbb{L}$, which arose according to possibility 1[8]:

An approximate image of the limit set according to Strategy 1

---

[8] The circles $D_{c_1,\ldots,c_k}$ have been colored.

Concluding remarks

1. Overall, a group $\mathcal{G}$ of the Schottky type is quite well understood due to our results. It is known how the $M \in \mathcal{G}$ act on the limit set $\mathbb{L}$ and on the complement, the union of the images of the fundamental domains (that is, the set $\bigcup_{T \in \mathcal{G}}$). And as a group, $\mathcal{G}$ is quite simple, it is the free group F_2.

2. Instead of starting with two pairs of circles, one can also specify k corresponding pairs of disjoint circles. The above results are easy to transfer. The group generated by the associated Möbius transformations is isomorphic to the free group with k generators. The limit set is always isomorphic to the Cantor discontinuum, and as a fundamental domain one can choose the complement of the $2k$ circles.

3. Here is a whole class of further examples. Let $0 < s < t < 1$. We consider the following four circles[9]:

$$K_1 := K_{(t+s)/2,(t-s)/2}, \quad K_2 := K_{-(t+s)/2,(t-s)/2},$$
$$K_1' := K_{(1/s-1/t)/2,(1/s+1/t)/2}, \quad K_1' := K_{-(1/s-1/t)/2,(1/s+1/t)/2}.$$

And these are the transformations by which the circles are "paired":

$$M_1 z = \frac{(s+t)z - 2st}{-2z + (s+t)}, \quad M_1' z = \frac{(s+t)z + 2}{2stz + (s+t)}.$$

The circles are therefore symmetrical to the real and imaginary axes, one can imagine them like this:

Another example of Schottky circles

And then you can look at the group $\mathcal{G}$ generated by M_1, M_1' again. This time, the images of the circles under the transformations of the group (after coloring) look like this:

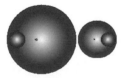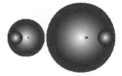

The pictures of the Schottky circles

---

[9] $K_{z,r}$ denotes the circle with center z and radius r

They always get smaller, and the circles contract to points on the real axis. The limit set is very similar to the Cantor discontinuum:

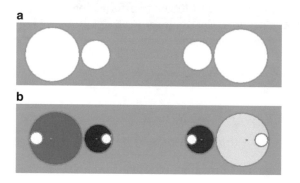

The limit set, approximately

And as a fundamental domain F we choose the complement of the four circles again. We see F and below the image of F under the first transformations:

Fundamental domain (**a**, *light green*) and the image under some transformations (**b**)

So nothing essentially changed as expected.

6.8   The Mystery of the Parabolic Commutator

So far we have considered *disjoint* Schottky circles. What happens if we enlarge them so that they just touch[10]? In principle, nothing changes, except that we can no longer guarantee that the radii of the image circles go to zero. Here is an example:

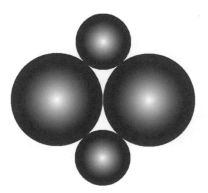

An example with touching circles

---

[10] In Mumford et al.'s book they are called "kissing Schottky disks".

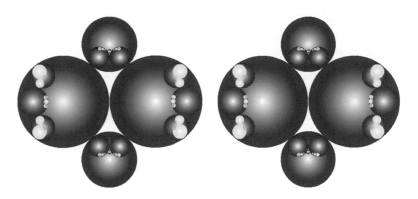

The first image circles and the limit set

But we would like to achieve that the limit set "grows together". What has to be done? We look at the following picture:

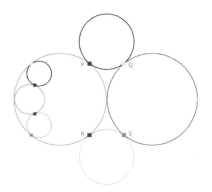

What has to be done that the limit set "grows together"?

With the notations used before:

- The transformations a and A pair the circles D_A (left) and D_a (right). a thus maps the interior resp. exterior of D_A to the exterior resp. interior of D_a.
- The transformations b and B pair the circles D_B (top) and D_b (bottom). b thus maps the interior resp. exterior of D_B to the exterior resp. interior of D_b.

We now analyze the effect of A on the circles D_A, D_b, D_B. (D_a is not mapped by A.) D_A and D_b touch at the point R (blue), and so it is not surprising that the image circles under A are tangent at the blue point. Similarly, D_A and D_B touch at the point P (red), and so it is clear that the image circles under A are tangent at the red point. Finally, it is clear that $A(Q)$ (yellow) and $A(S)$ (gray) are mapped to the green circle, because D_a and D_A are paired by a and A .

In order to arrive at a connected limit set, the circles $A(D_B)$ and $A(D_b)$ should touch D_b and D_B. This will be true if $A(Q) = P$ and $A(S) = R$. Similarly the conditions

$$a(P = Q), \ a(R) = S, \ B(R) = P, \ B(S) = Q, \ b(P) = R, \ b(Q) = S$$

would guarantee, that there are no gaps in the row of circles of the $D_{\alpha,\beta}$ (with $\alpha, \beta \in \{a, A, b, B\}$, without the α, β with $\alpha\beta = \mathrm{Id}$). This will then also be true for the $D_{\alpha,\beta,\gamma}$ etc., and therefore the limit set will be connected.

Some of the eight conditions can be omitted: $A(S) = P$, e.g., implies that $a(P) = S$ since $A = a^{-1}$. There remain 4 essential conditions. They have an interesting conseqence for the commutator of a and b:[11]

$$abAB(S) = abA(Q) = ab(P) = a(R) = S.$$

Thus S is a fixed point of $abAB$!

If you apply this equation to B, then $BabA\big(B(S)\big) = B(S)$ follows, i.e. Q is a fixed point of $BabA$, the commutator of B and a. Analogously, one shows: P is a fixed point of $ABab$, and R is a fixed point of $bABa$.

As an example, consider the point of contact of two circles, to which the circles D_a, $D_{a,b}$, $D_{a,b,A}$, $D_{a,b,A,B}$ come closer and closer from one side: The pattern a, b, A, B repeats itself over and over again. Therefore, the circles approach $z_{a,b,A,B,a,b,A,B,...}$, which is the attractive fixed point of $abAB$. And if $abAB$ were parabolic, that would be the only fixed point.

And from the other side, the circles D_b, $D_{b,a}$, $D_{b,a,B}$, $D_{b,a,B,A}$ have to be constructed: This time the pattern b, a, B, A repeats itself, and that is why these circles contract to the attracting fixed point of $baBA$. But $baBA = (abAB)^{-1}$, and therefore the second family of circles would converge towards the limit of the first if we assumed $abAB$ to be parabolic. This should be emphasized again:

- $\lim M^n z = \lim M^{-n} z$ for parabolic transformations M and every z.
- $baBA = (abAB)^{-1}$.
- So if $abAB$ is parabolic, the circles of higher order come together at the point where D_a and D_b touch.

By the way, no further conditions are needed, because then the images $BabA, ABab, bABa$ must also be parabolic: They arise from $abAB$ by conjugation. (For example, $BabA = B(abAB)B^{-1}$.)

---

[11] The *commutator* of two elements x, y of a group is the element $xyx^{-1}y^{-1}$. In a sense it measures to what extent x, y commute.

We formulate this as wishes for our *wish list*:

- The circles should touch each other, and the points of contact should be fixed points of the commutators.
- The commutators are parabolic.

If that is fulfilled, the circles really grow together. Here is an example[12]. First you see the four circles and the 12 circles $D_{\alpha,\beta}$ ($\alpha, \beta \in \{a, b, A, B\}$ and $\alpha\beta \neq \text{Id}$):

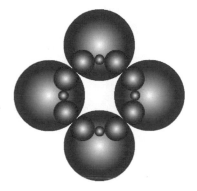

The four circles touching each other and the circles $D_{\alpha,\beta}$ with $\alpha\beta \neq \text{Id}$

And here are the $D_{\alpha,\beta,\gamma}$ and the $D_{\alpha,\beta,\gamma,\delta}$, with α, β, γ and $\alpha, \beta, \gamma, \delta$; nowhere are a transformation and its inverse next to each other:

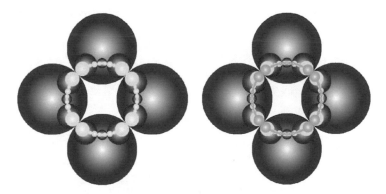

Here are the circles of the third and fourth generation:

---

[12] How our wish list can be fulfilled will be discussed in the next section.

Here is a detail:

A detail: Everything grows together due to the parabolic commutator.

One already suspects—and that is then also true—that the limit set is a circle. So the "dust" of the limit set in the previous section is now a connected set. Just as, for example, from the Cantor discontinuum (all numbers in $[0, 1]$, for which in the three-digit representation the 1 is missing) the whole interval is generated when one looks at *all* representations in the three-digit system.

There is still a subtlety to be observed. A parabolic Möbius transformation can still have trace is 2 or -2. Here we wish that the trace -2 is. Otherwise, the a, b, A, B are not suitable for our purposes. This is due to the following

Proposition 6.8.1

(i)   Let a, b, A, B be constructed according to the wish list. Then a and b have no common fixed point.

(ii) Let S, T be normalized Möbius transformations. If the trace of the commutator $STS^{-1}T^{-1}$ is equal to 2, then S and T have a common fixed point.

Proof (i) D_a touches D_b at the intersection of D_{ab} and D_b, and D_a touches D_B at the intersection of D_{aB} and D_B. Therefore, $D_{a,a}$ does not touch the circles D_b and D_B. The attracting fixed point of a is in D_{aa}, consequently not in D_b, where the attracting fixed point of b can be found. Analogously: The other fixed point of a is in $D_{A,A}$, and this circle does not meet D_b or D_B.

(ii) The relevant property (common fixed points, parabolic) is invariant under conjugation, so we only need to consider two cases.

Case 1: $Tz = \frac{az+b}{cz+d}$, and $Sz = \alpha z$. We assume both transformations to be normalized, i.e., $ad - bc = 1$, and S must be rewritten clumsily as $Sz = \gamma z/(1/\gamma)$, where γ is chosen so that $\gamma^2 = \alpha$. A longer calculation then shows that the trace of $STS^{-1}T^{-1}$ has the value $2 - bc\left(\gamma - \frac{1}{\gamma}\right)^2$.

If this trace is therefore equal to 2, $bc\left(\gamma - \frac{1}{\gamma}\right)^2 = 0$ must hold. This can only happen if $b = 0$ or $c = 0$ or $\alpha = \gamma^2 = 1$. In all cases, S and T have a common fixed point: If $b = 0$, then 0 is a common fixed point, if $c = 0$, then ∞ is a common fixed point, and in the third case $S = \text{Id}$.

Case 2: $Tz = \frac{az+b}{cz+d}$, and $Sz = z + \beta$. Again, a lot of calculation is required, precautions concerning S are not necessary this time. The trace of $STS^{-1}T^{-1}$ this time is $2 + \beta^2 c^2$. If the trace is equal to 2, $\beta^2 c^2 = 0$ must hold, and this implies $\beta = 0$ or $c = 0$. In the first case $S = \text{Id}$, in the second case ∞ is a common fixed point. $\quad\square$

As a consequence of this, we update our wish list:

- The circles should touch each other, and the points of contact should be fixed points of the commutators.
- The commutators are parabolic with trace -2.

Now we present a *first class of examples*, which we will take up again in the next section. Here is the "recipe[13]":

- Fix positive real numbers y, v with $yv > 1$ and set $x := \sqrt{1 + y^2}, u = \sqrt{1 + v^2}$. Determine a positive k with $0{,}5(k + 1/k) = 1/(yv)$.
- Set $a := \begin{pmatrix} u & ikv \\ -iv/k & u \end{pmatrix}$ and $b := \begin{pmatrix} x & y \\ y & x \end{pmatrix}$.

Lemma 6.8.2

(i)   These matrices generate normalized Möbius transformations.

(ii) In doing so, a pairs the circles with centers $\pm iku/v$ and radius k/v, and b pairs the circles with centers $\pm x/y$ and radius $1/y$. The circles are tangent to each other.

(iii) The commutator is parabolic with trace -2.

Proof (i) This is clear.

(ii) This follows quickly from the concrete construction. For example, to show that the circles touch, one must prove that the distance between the centers is equal to the sum of the radii, that is, e.g.

$$\left| \frac{iku}{v} - \frac{x}{y} \right| = \frac{k}{v} + \frac{1}{y}.$$

---

[13] I took it from the book by Mumford et al., p. 170.

This is, after squaring, equivalent to

$$\left(\frac{ku}{v}\right)^2 + \frac{x^2}{y^2} = \frac{1}{y^2} + \frac{2k}{yv} + \frac{k^2}{v^2}.$$

If you write $u^2 = 1 + v^2$ and $x^2 = 1 + y^2$, this reduces to the equation $k^2 + 1 = 2k/(yv)$ or $k + 1/k = 2/(yv)$. This is correct according to the assumption.

(iii) Later (in proposition 6.10.1) we will show that one only has to prove that $(Sp(a))^2 + (Sp(b))^2 + (Sp(ab))^2 = Sp(a)\,Sp(b)\,Sp(ab)$ is true. $(Sp(M))$ denotes the trace of a matrix. (In German "trace" is called "Spur", this explains the notation "Sp".) In the present case, one has to check whether

$$4u^2 + 4x^2 + \left(2ux + yvi(k - 1/k)\right)^2 = 4ux\left(2ux + yvi(k - 1/k)\right)$$

is true. This can be rewritten as $4 - y^2v^2(k - 1/k)^2 = 4v^2y^2$ and further to the equation $4 = y^2v^2\left(4 + (k - 1/k)^2\right)$. Now yv is replaced by $2/(k + 1/k)$ and $4(k + 1/k)^2 = 4\left(4 + (k - 1/k)^2\right)$ is obtained. This is easy to calculate.   □

Of course, one can also specify $k, y > 0$, the other numbers then are calculated as follows: $v = 2/(k + 1/k)$, $u = \sqrt{1 + v^2}$, $x = \sqrt{1 + y^2}$. Here you can see the Schottky circles and their images under "frequent" application of a, b, A, B (that is, an approximation of the limit set), first for $k = y = 1$:

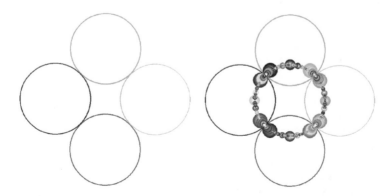

The circles and an approximation of the limit set in the case $k = y = 1$

The pictures are more interesting for other values of k, y, for example for $k = 0,2$ and $y = 1,5$:

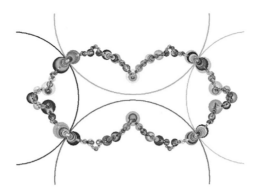

The approximation of the limit set in the case $k = 0,2$ and $y = 1,5$

The limit set looks like this in this case:

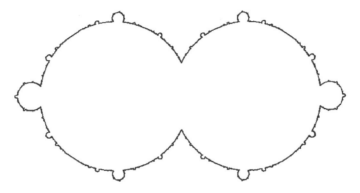

The limit set in the case $k = 0,2$ and $y = 1,5$

The examples show that the limit sets are somehow "fractal". But it is still a closed curve, which is why the resulting groups belong to the class of generalized *Fuchsian groups*.

Below you can see the resulting limit sets when taking a "walk" through the parameter space of k, y:

The limit sets for $k = 1, k = 0,8$ and $k = 0,6$; here always $y = 1,4$

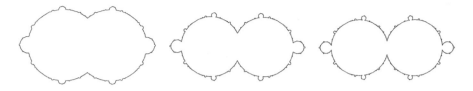

The limit sets for $k = 0,4$, $k = 0,2$ and $k = 0,1$; again we always put $y = 1,4$

6.9   The Structure of Kleinian Groups

We have investigated some classes of Kleinian groups $\mathcal{G}$. But what general properties are there, how can one imagine the transformations induced by such a group?

It will be shown that there is a *fundamental domain* $\mathbb{F}$: The $M(\mathbb{F})$ with $M \in \mathcal{G}$
tessellate $\hat{\mathbb{C}}$ "except for a small exceptional set". It can have a very interesting
fractal structure, which we will examine in more detail in the next section. On the
limit set $\mathcal{G}$ operates quite chaotically.

We will proceed as follows: After a general preparation, we examine special
circles that play an important role here: the *isometric circles* of Möbius transformations. The totality of these circles has very special properties in the case of a
Kleinian group, which in particular motivate us to define the limit set and a candidate for a fundamental domain.

First a short preparation. We fix an arbitrary Kleinian group $\mathcal{G}$. Then there is
a z_0 and a neighbourhood U of z_0, so that Mz_0 for all non-trivial $M \in \mathcal{G}$ is not in
U. We know that for structural investigations a subgroup can be arbitrarily conjugated, so we can pass from $\mathcal{G}$ to $\{GMG^{-1} \mid M \in \mathcal{G}\}$, where G is an arbitrary
Möbius transformation. The essential group properties remain unchanged. We may
and will w.l.o.g assume that $z_0 = \infty$ is true, for this one must only choose G such
that $Gz_0 = \infty$. Then it follows:

- There is an $R > 0$, so that $|M(\infty)| \leq R$ for all $M \in \mathcal{G} \setminus \{\text{Id}\}$. This means: If
  $M = M_{a,b,c,d} \in \mathcal{G} \setminus \{\text{Id}\}$, then $|a/c| \leq R$. In particular, $c \neq 0$ is always true for
  these M.
- ∞ is not a fixed point for the $M \in \mathcal{G} \setminus \{\text{Id}\}$.

This should always be satisfied for the following investigations.

6.9.1   The Isometric Circles

Some circles have a special meaning for a Möbius transformation M. We already
know: If $f : \mathbb{C} \to \mathbb{C}$ is holomorphic, then f is for a z_0 for "small" w well approximated by

$$f(z_0 + w) \approx f(z_0) + f'(z_0)w$$

Consequently, the absolute size of vectors (or more generally a picture) is preserved locally if $|f'(z_0)| = 1$.

In particular, for (normalized) Möbius transformations $M = M_{a,b,c,d}$ this means:

- If $c = 0$, then $M(z) = (a/d)z + b/d$. Therefore, M' is the constant vector a/d, and thus M is isometric at all z (if $|a/d| = 1$) or at none (if $|a/d| \neq 1$).
- If $c \neq 0$, then

$$M'(z) = \frac{(cz+d)a - c(az+b)}{(cz+d)^2} = \frac{1}{(cz+d)^2}.$$

M is therefore isometric locally at z when $|cz + d| = 1$. This is a circle around $-d/c$ with radius $1/|c|$. It is called *the isometric circle associated with M*. We denote it by I_M, and instead of $I_{(M^{-1})}$ we will write more briefly I'_M. The radius is always positive in the case $c \neq 0$.

Lemma 6.9.1
Let $M = M_{a,b,c,d}$ be normalized, and $c \neq 0$.

(i) The singularity $-d/c$ of M is the center of I_M. Points inside I_M are stretched locally, points outside are compressed locally.
(ii)   For the circles I_M and I'_M one has: Both circles have the radius $1/|c|$, and $M(I_M) = I'_M$.

Proof (i) This is clear, because the local magnification factor is $1/|cz + d|^2$.

(ii) Because of Lemma 5.2.2 we know that $M^{-1} = M_{d,-b,-c,a}$, I'_M is therefore the circle with the radius $1/|c|$ and the center a/c. The claim thus amounts to inferring $|cz + d| = 1$ from $|cM(z) - a| = 1$. For this we use the formula known from analysis $(M^{-1})'(M(z)) = 1/M'(z)$, it implies $1/(cM(z) - a)^2 = (cz + d)^2$ and thus the claim. It should be clear that the statement also follows from the fact that M is isometric on I_M. Then M^{-1} must also be isometric on I'_M.   □

I_M and I'_M can be very different from each other: They can be identical, they can intersect or be disjoint. For $Mz = M_{0,1,1,0}z = 1/z$ is $I_M = I'_M$ the unit circle, in the case $M_{0,1,1,1}$ the circles intersect, and $M_{0,1,1,2,1}$ leads to disjoint isometric circles.

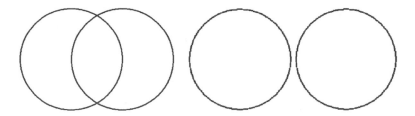

I_M (*red*) and I'_M (*blue*) for $M_{0,1,1,1}$ and $M_{0,1,1,2,1}$

It will be very important to imagine the effect of M on these circles. M maps the boundary of I_M to the boundary of I'_M, and the the center of I_M is mapped to ∞. The image of the circle I_M under M must therefore be the exterior of $I_{M'}$, and the image of the exterior of I_M is necessarily the interior of $I_{M'}$. Accordingly, M^{-1} maps the interior (or exterior) of I'_M onto the exterior (or interior) of I_M. The situation is thus similar to the case of paired Schottky circles, which we could specify arbitrarily.

So far we have defined the I_M for individual M. Now we want to investigate what happens when we consider all M from our Kleinian group $\mathcal{G}$. (Note that we consider a $\mathcal{G}$ where always $c \neq 0$, the I_M are thus defined for all $M \neq \mathrm{Id}$.) As a preparation, we calculate how the isometric circles behave when composing the associated images.

Let $M = M_{\alpha,\beta,\gamma,\delta}$ and $N = M_{a,b,c,d}$ be normalized Möbius transformations. We assume that the isometric circle is defined for N, M and MN, i. e. $c \neq 0 \neq \gamma$ and $\gamma a + \delta c \neq 0$. In particular, $M \neq N^{-1}$.

Lemma 6.9.2

The centers of I_M or I'_M or I_N or I'_N or I_{MN} will be called g_M or g'_M or g_N or g'_N or g_{MN} and the radii of I_M or I_N or I_{MN} will be denoted by r_M or r_N or r_{MN}. Then the following is true:

(i) $r_{MN} = r_M r_N / |g'_N - g_M|$.
(ii) $|g_{MN} - g_N| = r_N^2 / |g'_N - g_M|$.
(iii) For all centers g, one has $|g| \leq R$. And every I_M lies in the circle around the origin with the radius $3R$.

Proof (i) Explicitly we get, after calculation of MN,

$$r_M = \frac{1}{|\gamma|}, \quad r_N = \frac{1}{|c|}, \quad r_{MN} = \frac{1}{|\gamma a + \delta c|}$$

as well as

$$g_M = -\frac{\delta}{\gamma}, \quad g'_M = \frac{\alpha}{\gamma}, \quad g_N = -\frac{d}{c}, \quad g'_N = \frac{a}{c}, \quad g_{MN} = -\frac{\gamma b + \delta d}{\gamma a + \delta c}.$$

It follows that

$$r_{MN} = \frac{1}{|\gamma a + \delta c|}$$
$$= \frac{1}{|\gamma c| \cdot |a/c + \delta/\gamma|}$$
$$= \frac{r_M r_N}{|g'_N - g_M|}.$$

(ii) Due to the above formulas,

$$g_{MN} - g_N = -\frac{\gamma b + \delta d}{\gamma a + \delta c} + \frac{d}{c} = \frac{\gamma}{c(\gamma a + \delta c)}.$$

So it follows

$$|g_{MN} - g_N| = \frac{r_{MN} r_N}{r_M} = \frac{r_N^2}{|g_N' - g_M|}.$$

(iii) Because of $g_M = M^{-1}(\infty), |g_M| \le R$. So it follows

$$r_N^2 = |g_{MN} - g_N| \, |g_N' - g_M| \le 4R^2,$$

and therefore $r_N \le 2R$ for arbitrary N.   □

The lemma has far-reaching consequences for the isometric circles, which belong to the transformations of our Kleinian group $\mathcal{G}$:

Proposition 6.9.3
As agreed at the beginning, we assume that $|M(\infty)| \le R$ for a suitable R and all non-trivial $M \in \mathcal{G}$.

(i)   For all $M \in \mathcal{G}$ with $M \ne \mathrm{Id}$, $r_M \le 2R$ and $|g_M| \le R$.
(ii) For $M, N \in \mathcal{G} \setminus \{\mathrm{Id}\}$ and $M \ne N$, $g_M \ne g_N$. It follows: With $\mathcal{G}$, the set $\{g_M \mid M \in \mathcal{G}, M \ne \mathrm{Id}\}$ of centers of isometric circles is also infinite.
(iii) For each $\varepsilon > 0$, there are only finitely many $M \in \mathcal{G} \setminus \{\mathrm{Id}\}$ with $r_M > \varepsilon$.

Proof (i) Let $\tilde{M} \in \mathcal{G}$. Then $\tilde{M}(g_{\tilde{M}}) = \infty$, so $\tilde{M}^{-1}(\infty) = g_{\tilde{M}}$. Consequently, $|g_{\tilde{M}}| \le R$ for all $\tilde{M} \in \mathcal{G}$. From the previous lemma we deduce for $M \in \mathcal{G} \setminus \{\mathrm{Id}\}$:

$$r_M^2 = |g_{MN} - g_M| \, |g_N' - g_M| \le 4R^2,$$

i.e. $r_M \le 2R$.

(ii) If $M \ne N$, then $MN^{-1} \ne \mathrm{Id}$, so we can apply the previous lemma to the transformations M, N^{-1}. It follows that

$$|g_N - g_M| = (r_N')^2 / |g_{MN^{-1}} - g_N'| > 0.$$

(iii) Let $\varepsilon > 0$, and $I_M, I_{N'}$ be different isometric circles with $r_M, r_N' > \varepsilon$. Then $MN \ne \mathrm{Id}$ (otherwise $I_M = I_N'$) and consequently

$$|g_N' - g_M| = \frac{r_M r_N}{r_{MN}} > \frac{\varepsilon^2}{2R}.$$

g'_N, g_M lie in the circle with radius R, and therefore the number must be finite. □

As a consequence it follows:

> **Corollary 6.9.4**
> If $I_1, I_2, \ldots$ is a sequence of pairwise different isometric circles of elements
> from $\mathcal{G}$, then the radii go to zero.

6.9.2   The Limit Set

As before, $\mathcal{G}$ is a Kleinian group for which the values of $M(\infty)$ (with $M \in \mathcal{G}$,
$M \neq \text{Id}$) are bounded. We call a z_0 a *limit point* of $\mathcal{G}$ if z_0 is the accumulation point
of the centers of the isometric circles of the $M \in \mathcal{G}$[14]. $\mathbb{L}$, the *limit set of $\mathcal{G}$*, denotes
the set of limit points. (Actually, one should say "limit point with respect to $\mathcal{G}$" or
something similar and write $\mathbb{L}_\mathcal{G}$. Since $\mathcal{G}$ is fixed, however, there is no risk of con-
fusion.)

In the following picture, "many" centers of isocircles are drawn for a Schottky
group with parabolic commutator. There are some isolated points, and one can see
that $\mathbb{L}$ consists of the points on the circle:

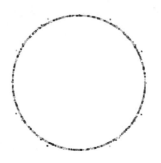

Centers of isocircles and accumulation points

$\mathbb{L}$ is always closed and in the case of finite groups empty. For infinite sets, it follows from proposition 6.9.3 (ii) that there are limit points. In the examples already
treated, we have:

---

[14] z_0 is called an *accumulation point* of a subset D of $\mathbb{C}$ if for each $\varepsilon > 0$ there is an $z \in D$ with
$0 < |z_0 - z| \leq \varepsilon$. The set of accumulation points of a set is always closed.

- $\mathbb{L}$ can be one-point (e.g. for the cyclic group generated by a parabolic transformation).
- $\mathbb{L}$ can be two-point (e.g. for the cyclic group generated by a hyperbolic transformation).
- $\mathbb{L}$ can be uncountable (e.g. in the case of Schottky groups).

(These statements are easy to see. In the case of Schottky groups, we had defined limit points differently, but the definition is equivalent.)

The systematic investigation of $\mathbb{L}$ begins with the

Lemma 6.9.5

$z_0, z_1, z_2 \in \mathbb{C}$ are pairwise different, and $z_0 \in \mathbb{L}$. Then there is to each $\varepsilon > 0$ an $\tilde{M} \in \mathcal{G}$, so that $0 < |z_0 - \tilde{M}z_1| \leq \varepsilon$ or $0 < |z_0 - \tilde{M}z_2| \leq \varepsilon$.

Proof Let $\varepsilon > 0$ be given, w.l.o.g. let all distances $|z_i - z_j|$ $(i \neq j)$ be greater than 2ε. Choose a pair of different centers of isometric circles g_n with $g_n \to z_0$. Because of Corollary 6.9.4 the radii go to zero, we find an $M \in \mathcal{G}$, for which the isometric circle is entirely contained in the circle around z_0 with the radius ε. In particular, $r_M \leq \varepsilon$.

Also I'_M has radius r_M, so due to our assumption z_1, z_2 cannot both be in I'_M. Let, for example, $z_1 \notin I'_M$. Then z_1 is mapped under $N := M^{-1}$ to I_M, one has therefore $|Nz_1 - z_0| \leq \varepsilon$.

If $Nz_1 \neq z_0$, then the statement is proved with $\tilde{M} = N$. But it could be that $Nz_1 = z_0$. z_0 is not in $I'_M = I_N$, the point z_0 is therefore mapped under N to $I'_N = I_M$, and therefore $|Nz_0 - z_0| \leq \varepsilon$. Because of $Nz_1 = z_0$ z_0 cannot be a fixed point of N, it therefore even follows that $0 < |Nz_0 - z_0| = |N^2z_1 - z_0| \leq \varepsilon$, the claim is therefore shown with $\tilde{M} = N^2$. $\qquad\square$

We assume that $\mathcal{G}$ contains infinitely many elements and $\mathbb{L}$ is therefore not empty.

Proposition 6.9.6

(i)   If $M \in \mathcal{G}$, then $g_M \notin \mathbb{L}$.

(ii) $\mathbb{L}$ is closed and invariant under all $M \in \mathcal{G}$.

(iii)   For each $M \in \mathcal{G}$, the distance from g_M to $\mathbb{L}$ is strictly positive. It follows that $\mathbb{L}$ has no interior.

(iv)   Suppose that $K \subset \mathbb{C}$ is closed and invariant under all $M \in \mathcal{G}$. If K contains at least two elements, then $\mathbb{L} \subset K$.

(v)   If $\mathbb{L}$ contains more than two elements, then $\mathbb{L}$ is a perfect set, each $z \in \mathbb{L}$ therefore lies in the closure of $\mathbb{L} \setminus \{z\}$. In particular, $\mathbb{L}$ is uncountable.

Proof (i) Otherwise, there would be pairs of different M_n with $M_n^{-1}(\infty) = g_{M_n} \to g_M$. The map M is singular at g_M, it follows that $M(M_n^{-1}(\infty)) \to \infty$. This contradicts the fact that the $MM_n^{-1}(\infty)$ are bounded by R.

(ii) Let $z_0 \in \mathbb{L}$ and $M \in \mathcal{G}$. According to the first part, M is continuous at z_0. Choose pairs of different centers of isometric circles g_n with $g_n \to z_0$. We write $g_n = M_n(\infty)$ for suitable $M_n \in \mathcal{G}$.

Then $M(g_n) = MM_n(\infty)$, and these numbers converge to Mz_0. Since the MM_n are pairwise different, this proves $Mz_0 \in \mathbb{L}$.

(iii) This follows from the two previous statements: Centers of isometric circles come arbitrarily close to the z_0, but – together with a neighborhood – do not belong to it.

(iv) Let z_1, z_2 be two different points from K and $z_0 \in \mathbb{L}$. If $z_0 \notin K$, one could choose a $\varepsilon > 0$ so that $|z - z_0| > \varepsilon$ for all $z \in K$. We apply the above lemma to z_0, z_1, z_2. This gives us an $M \in \mathcal{G}$, so that $|Mz_1 - z_0| \leq \varepsilon$ or $|Mz_2 - z_0| \leq \varepsilon$. One would therefore have $Mz_1 \notin K$ or $Mz_2 \notin K$. This contradicts the invariance of K under the $M \in \mathcal{G}$.

(v) Let $z_0 \in \mathbb{L}$. We choose $z_1, z_2 \in \mathbb{L}$ and apply the above lemma several times. We start with an $\varepsilon_0 > 0$. For a suitable $M_1 \in \mathcal{G}$, $0 < |M_1 z_1 - z_0| \leq \varepsilon_0$ or $0 < |M_1 z_2 - z_0| \leq \varepsilon_0$. Let, for example, $0 < |M_1 z_1 - z_0| \leq \varepsilon_0$. Set $\varepsilon_1 := |M_1 z_1 - z_0|/2$ and apply the lemma again. So $M_1, M_2, \ldots$ arise, for which $M_n z_1$ or $M_n z_2$ approaches z_0. Because of part (i), the $M_n z_1$ or $M_n z_2$ belong to $\mathbb{L}$, i. e., z_0 is an accumulation point of elements of $\mathbb{L}$.

We still sketch why $\mathbb{L}$ must then be uncountable. Start with $z_0 \neq z_1$ in $\mathbb{L}$ and choose disjoint closed ε_1-neighborhoods U_0, U_1. Choose in U_0 two different z_{00}, z_{01} from $\mathbb{L}$ and disjoint closed ε_2-neighborhoods $U_{00}, U_{01} \subset U_0$; correspondingly, z_{10}, z_{11} and neighborhoods $U_{10}, U_{11} \subset U_1$ are chosen.

This is continued, there are non-empty closed sets $U_{i_1 i_2 \ldots i_k}$ where $k \in \mathbb{N}$ and $i_1, \ldots, i_k \in \{0, 1\}$. Map $\{0, 1\}^{\mathbb{N}}$ to $\mathbb{L}$ that $i_1 i_2 \ldots$ is mapped to the uniquely determined element in the intersection of the $U_{i_1 \ldots i_k}, k \in \mathbb{N}$. $\mathbb{L}$ is closed, the intersection really contains only one element according to the Cantor intersection theorem, and the mapping is injective. Summing up, this means: $\mathbb{L}$ contains at least as many elements as $\{0, 1\}^{\mathbb{N}}$, and these are uncountably many. $\square$

We still want to explain why the definition given here for limit sets in the case of Schottky groups agrees with the one given in Sect. 6.7. For this one must remember that there are no elliptic transformations in Schottky groups and then combine this fact with the following proposition:

Proposition 6.9.7
Let $\mathcal{G}$ be discrete and $M \in \mathcal{G}$ an element that is not elliptic. Then the fixed points of M belong to $\mathbb{L}$.

Proof The statement is invariant under conjugation, and therefore it is enough to show the statement for transformations M which are conjugate to $z \mapsto z + \beta$ or $z \mapsto \alpha z$. We conjugate with $G = \left(\begin{smallmatrix} w & -w \\ w & w \end{smallmatrix} \right)$ where $w := 1/\sqrt{2}$. Then $G^{-1} = \left(\begin{smallmatrix} w & w \\ -w & w \end{smallmatrix} \right)$.

Case 1: $Mz = z + \beta$ for a $\beta \neq 0$. A short calculation shows that

$$GM^nG^{-1} = \begin{pmatrix} 1 - \frac{n\beta}{2} & \frac{n\beta}{2} \\ -\frac{n\beta}{2} & 1 + \frac{n\beta}{2} \end{pmatrix}.$$

All GM^nG^{-1} are therefore different, and the isometric circle of this transformation has the center $1 - 2/(n\beta)$. These numbers converge to 1, and this number is also the fixed point of GMG^{-1}.

Case 2: $Mz = \alpha z$ for a α with $|\alpha| \neq 1$. We write M normalized as $(\eta z/(1/\eta))$ for an η with $\eta^2 = \alpha$ and conjugate M^n again with G:

$$G \begin{pmatrix} \eta^n & 0 \\ 0 & 1/\eta^n \end{pmatrix} G^{-1} = \begin{pmatrix} (\eta^n + 1/\eta^n)/2 & (\eta^n - 1/\eta^n)/2 \\ (\eta^n - 1/\eta^n)/2 & (\eta^n + 1/\eta^n)/2 \end{pmatrix}.$$

The isometric circle of the conjugated M^n therefore has radius $2/|\eta^n - 1/\eta^n|$ and center $(\eta^n - 1/\eta^n)/(\eta^n + 1/\eta^n)$. For $n \to \pm\infty$ these centers go to ± 1, the fixed points of GMG^{-1}. □

Now we can also justify why the condition in part (ii) of the proposition that K has at least two elements cannot be omitted. Just think of the cyclic group generated by a hyperbolic transformation M. The fixed points of M are ξ_1 and ξ_2. Then $\mathbb{L} = \{\xi_1, \xi_2\}$, and $K := \{\xi_1\}$ are invariant under all group elements $M^n, n \in \mathbb{Z}$.

Also, in (iii) it is important that $\mathbb{L}$ contains more than two elements. $\mathbb{L}$ can be a one-point or two-point set, as the example of cyclic groups of parabolic and hyperbolic transformations shows. And then $\mathbb{L}$ is of course not perfect.

6.9.3   A Fundamental Domain

Let $\mathcal{G}$ be given as above: $\mathcal{G}$ is discrete, and the $|M(\infty)|$ for $M \in \mathcal{G} \setminus \{\mathrm{Id}\}$ are limited by a number R. How could one now find a fundamental domain, that is, a subset $\mathbb{F}$ of $\hat{\mathbb{C}}$ for which the $M(\mathbb{F})$ with $M \in \mathcal{G}$ tesselate the set $\hat{\mathbb{C}}$ "as well as possible". Our aim will be modest: We want to find $\mathbb{F}$ so that the $M(\mathbb{F})$ with $M \in \mathcal{G}$ are first pairwise disjoint and secondly their union is dense[15].

We begin by asking the question: How can one find a $z \in \hat{\mathbb{C}}$ that all Mz with $M \in \mathcal{G}$ are different? Let's assume there is an $M \in \mathcal{G} \setminus \{\mathrm{Id}\}$ so that z is neither in the isometric circle I_M of M nor in the circle I'_M of M^{-1}. Then Mz (or $M^{-1}z$) is

---

[15] How to achieve more is described in Ford's book

mapped to I'_M (or I_M), so it is certainly not mapped to z. If one wants to ensure this for all M, then z must be outside all isometric circles. This motivates the following

---

Definition 6.9.8
We denote by $\mathbb{F}$ the complement of the union of all I_M for the $M \in \mathcal{G} \setminus \{\text{Id}\}$.

From the above considerations it is then clear that $\mathbb{F}$ is disjoint with all $M(\mathbb{F})$ with $M \neq \text{Id}$, and thus also $M(\mathbb{F})$ and $N(\mathbb{F})$ have no common points for $M \neq N$. In the case of Schottky groups, $\bigcup I_M$ is actually the union of finitely many disks, but the $\mathbb{F}$ defined here does not have to coincide with the one treated in Schottky groups. $\mathbb{F}$ is also large: Since all I_M are in $\{|z| \leq 3R\}$, $\mathbb{F}$ contains all z with $|z| > 3R$.

Here is our main result:

Proposition 6.9.9
$\mathbb{F}$ is a fundamental domain for $\mathcal{G}$ in the following sense: The $M(\mathbb{F})$ are pairwise disjoint, and $\bigcup_{M \in \mathcal{G}} M(\mathbb{F})$ is dense in $\hat{\mathbb{C}}$.

Proof Assume that the closure of the set $\bigcup_{M \in \mathcal{G}} M(\mathbb{F})$ would not contain a z_0. Because of $\infty \in \mathbb{F}$, this means $z_0 \in \mathbb{C}$. Choose a circular disk Q with center z_0 and radius $r > 0$, so that Q and $\bigcup_{M \in \mathcal{G}} M(\mathbb{F})$ have no common points. Of course, then $N(Q)$ and $\bigcup_{M \in \mathcal{G}} M(\mathbb{F})$ are also disjoint. Our aim: Among the $N(Q)$ there are circles with arbitrarily large radius, and that contradicts the fact that $\mathbb{F}$ contains all z with $|z| > 3R$.

First we notice that no g_M is in Q, because $g_M = M^{-1}(\infty)$, and $\infty \in \mathbb{F}$. (It follows that no limit point can be inside Q.) On the other hand, $z_0 \notin \mathbb{F}$, the number z_0 therefore lies in a circle I_1 belonging to a transformation M_1 with radius r_1 and center g_1. The situation is therefore as shown in the following picture:

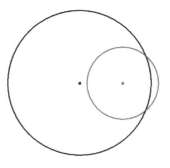

The circle I_M (*blue*) and the circle Q (*gray*)

One has $r < |g_1 - z_0| \leq r_1$. What happens now if you apply M_1 to Q? First, $Q_2 := M_1(Q)$ is a circle again, and secondly, Q_1 is disjoint from $\bigcup_{M \in \mathcal{G}} M(\mathbb{F})$. What happens to the radii? As an example for illustration, we consider $M_1 = 1/z$. There I_1 is the unit circle, and as Q we want to imagine a circle with center $z_0 \in\ ]0, 1]$ and radius $r < z_0$. The image circle then passes through the points $1/(z_0 \pm r)$, so it has radius $r/(z_0^2 - r^2) \geq r/(1 - r^2)$.

In the general case it follows: The radius of Q_2 is bounded from below by

$$r_2 := \frac{rr_1^2}{r_1^2 - r^2} = \frac{r}{1 - (r^2/r_1^2)}.$$

Here $r_1 \leq 2R$, so $r_2 \geq kr$, where $k := 1/(1 - r^2/4R^2) > 1$. If you apply this idea again, you will find an M_2, so that $U_3 := M_2 U_2$ is first disjoint from $\bigcup_{M \in \mathcal{G}} M(\mathbb{F})$ and secondly has a radius of at least $kr_1 \geq k^2 r$. In the n-th step, a circle disjoint from $\bigcup_{M \in \mathcal{G}} M(\mathbb{F})$ with radius (at least) $k^n r$ is then found. That can't be, because $k^n r$ will be arbitrarily large. This contradiction proves the claim.   □

Three more examples follow. First we consider the finite group generated by an elliptic transformation of order 6. It is transformed such that ∞ is not a fixed point. Here you can see the isometric circles:

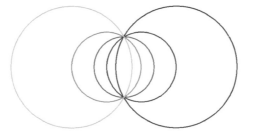

Isometric circles for a cyclic group with 6 elements

The union of *all* circles is already realized by the union of 2 circles. The complement is our fundamental domain $\mathbb{F}$. We see it in the following picture (red) together with the $M(\mathbb{F})$ for the non-trivial M of the group:

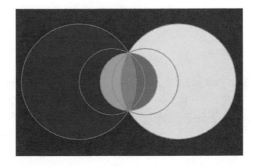

The corresponding fundamental domain (*red*)

An example of a non-commutative group follows, the so-called *anharmonic group*. It consists of the transformations

$$\text{Id}, \quad z, \quad \frac{1}{z}, \quad 1-z, \quad \frac{1}{1-z}, \quad \frac{z-1}{z}, \quad \frac{z}{z-1},$$

but it was conjugated to have ∞ not as a fixed point. In the following image you can see the isometric circles and the fundamental domain $\mathbb{F}$:

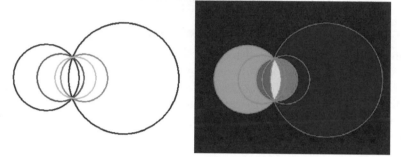

Isometric circles and fundamental domain (*red*) for the anharmonic group

The pictures of both groups are not distinguishable, the algebraic structure (commutative, not commutative) is not encrypted in the fundamental domain.

And here is another example of an infinite non-commutative group. We consider the Schottky group associated with the following circle pairing:

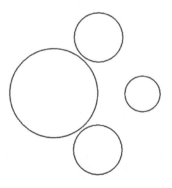

The circles of a Schottky group

In the next picture you can see the isometric circles to the Schottky circles. Sometimes they are different from the Schottky circles. (This is also to be expected if the paired circles are of different size: Isometric circles to M and M^{-1} always have the same radius.)

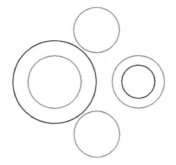

The circles of a Schottky group and the isometric circles of the associated transformations

Next, in addition isometric circles were also calculated for all M_1M_2 (left) and all $M_1M_2M_3$ (right):

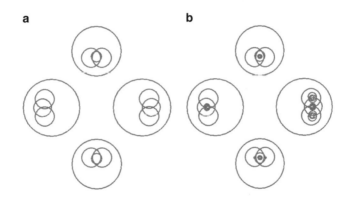

The isometric circles of a Schottky group for the M_1M_2 (**a**) and the $M_1M_2M_3$ (**b**)

Finally, a fundamental domain can be found: The complement of the union of the isometric circles (in the following image on the left). For comparison, another example of a fundamental domain is shown (on the right): The complement of the union of the Schottky circles.

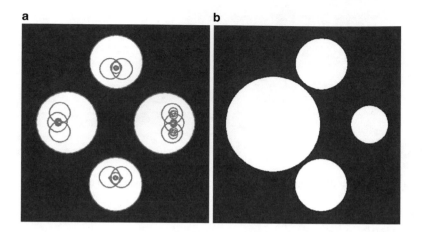

Fundamental areas: complement of isometric circles (**a**) and complement of Schottky circles (**b**)

6.10 Parabolic Commutators: Construction

How do you get transformations for which the commutator has trace -2? To do this, it will be helpful to first derive some formulas for traces. For a 2×2-matrix or a Möbius transformation M, we will use $\mathrm{Sp}(M)$ to denote the *trace of M*. Since the transition matrix $\to$ transformation is a group morphism, all equations that have been proven for matrices can be transferred to transformations.

Proposition 6.10.1
Let M, N, a, b be normalized 2×2-matrices, $A := a^{-1}, B := b^{-1}$.

(i) $\mathrm{Sp}(MN) + \mathrm{Sp}(M^{-1}N) = \mathrm{Sp}(M)\,\mathrm{Sp}(N)$.

(ii) $\mathrm{Sp}(abAB) = (\mathrm{Sp}(a))^2 + (\mathrm{Sp}(b))^2 + (\mathrm{Sp}(ab))^2 - \mathrm{Sp}(a)\,\mathrm{Sp}(b)\,\mathrm{Sp}(ab) - 2$
(Markov identity).

(iii) One has $\mathrm{Sp}(abAB) = -2$, exactly when $(\mathrm{Sp}(a))^2 + (\mathrm{Sp}(b))^2 + (\mathrm{Sp}(ab))^2 - \mathrm{Sp}(a)\,\mathrm{Sp}(b)\,\mathrm{Sp}(ab) = 0$, that is, when given $\mathrm{Sp}(a), \mathrm{Sp}(b)$, the trace of ab satisfies a certain quadratic equation.

Proof (i) This can be calculated directly. Write $M = \begin{pmatrix} e & f \\ g & h \end{pmatrix}$ and $N = \begin{pmatrix} i & j \\ k & l \end{pmatrix}$. Then $M^{-1} = \begin{pmatrix} h & -f \\ -g & e \end{pmatrix}$, and it follows

$$\mathrm{Sp}(MN) + \mathrm{Sp}(M^{-1}N) = (ei + fk) + (gj + hl) + (hi - fk) + (-gj + el).$$

This agrees with $\mathrm{Sp}(M)\,\mathrm{Sp}(N) = (e+h)(i+l)$.

(ii) We use the above formula for different M, N:

(1) For $M = a, N = b$ it follows $\mathrm{Sp}(ab) + \mathrm{Sp}(Ab) = \mathrm{Sp}(a)\,\mathrm{Sp}(b)$.

(2) For $M = a, N = bAB$ it follows $\mathrm{Sp}(abAB) + \mathrm{Sp}(AbAB) = \mathrm{Sp}(a)\,\mathrm{Sp}(bAB)$.

(3) For $M = Ab, N = AB$ it follows $\mathrm{Sp}(AbAB) + \mathrm{Sp}(B^2) = \mathrm{Sp}(Ab)\,\mathrm{Sp}(AB)$.

(4) For $M = N = B$ it follows $\mathrm{Sp}(B^2) + \mathrm{Sp}(\mathrm{Id}) = (\mathrm{Sp}(B))^2$.

We combine this with the known equations $\mathrm{Sp}(MN) \overset{(5)}{=} \mathrm{Sp}(NM)$, $\mathrm{Sp}(M) \overset{(6)}{=} \mathrm{Sp}(NMN^{-1})$; we also note that $\mathrm{Sp}(M) \overset{(7)}{=} \mathrm{Sp}(M^{-1})$ for normalized 2×2 matrices, so that $\mathrm{Sp}(AB) \overset{(8)}{=} \mathrm{Sp}(ab)$. It follows

$$
\begin{aligned}
\mathrm{Sp}(abAB) &\overset{(2)}{=} \mathrm{Sp}(a)\,\mathrm{Sp}(bAB) - \mathrm{Sp}(AbAB) \\
&\overset{(6)}{=} \mathrm{Sp}(a)\,\mathrm{Sp}(A) - \mathrm{Sp}(AbAB) \\
&\overset{(7)}{=} (\mathrm{Sp}(a))^2 - \mathrm{Sp}(AbAB) \\
&\overset{(3)}{=} (\mathrm{Sp}(a))^2 + \mathrm{Sp}(B^2) - \mathrm{Sp}(Ab)\,\mathrm{Sp}(AB) \\
&\overset{(4)}{=} (\mathrm{Sp}(a))^2 + (\mathrm{Sp}(B))^2 - \mathrm{Sp}(\mathrm{Id}) - \mathrm{Sp}(Ab)\,\mathrm{Sp}(AB) \\
&\overset{(7)}{=} (\mathrm{Sp}(a))^2 + (\mathrm{Sp}(b))^2 - 2 - \mathrm{Sp}(Ab)\,\mathrm{Sp}(AB) \\
&\overset{(1)}{=} (\mathrm{Sp}(a))^2 + (\mathrm{Sp}(b))^2 - 2 - (\mathrm{Sp}(a)\,\mathrm{Sp}(b) - \mathrm{Sp}(ab))\,\mathrm{Sp}(AB) \\
&\overset{(8)}{=} (\mathrm{Sp}(a))^2 + (\mathrm{Sp}(b))^2 - 2 - (\mathrm{Sp}(a)\,\mathrm{Sp}(b) - \mathrm{Sp}(ab))\,\mathrm{Sp}(ab) \\
&= (\mathrm{Sp}(a))^2 + (\mathrm{Sp}(b))^2 + (\mathrm{Sp}(ab))^2 - \mathrm{Sp}(a)\,\mathrm{Sp}(b)\,\mathrm{Sp}(ab) - 2.
\end{aligned}
$$

(iii) This follows immediately from (ii).   □

The trace equation makes it much easier to find transformations a, b for which the commutator is parabolic with trace -2^{16}.

The Problem Assume we want to find suitable matrices a, b for which the traces have given values $t_a, t_b \in \mathbb{C}$. In a first step, then t_{ab}, the trace of ab, must be chosen correctly: Due to the Markov equation, t_{ab} must be a solution of the quadratic equation

$$
x^2 - t_a t_b x + t_a^2 + t_b^2 = 0.
$$

So such a t_{ab} will always exist, and in general there are even two solutions. Now we make the very general ansatz

$$
a = \begin{pmatrix} A & B \\ C & D \end{pmatrix}, \quad b = \begin{pmatrix} E & F \\ G & H \end{pmatrix}
$$

---

[16] However, this does not yet answer the question of whether a and b also generate a *discrete* group and whether one can therefore expect an interesting limit set. This must always be checked separately in each individual case.

with $A, B, C, D, E, F, G, H \in \mathbb{C}$. To guarantee that the traces and determinants of these matrices have the right value,

$$AD - BC = EH - FG = 1, \quad t_a = A + D, \quad t_b = E + H$$

must hold. These are four (linear and quadratic) conditions in $\mathbb{C}^8$, so in competition is still a four-dimensional hypersurface of elements (A, B, C, D, E, F, G, H) in $\mathbb{C}^8$. With these a, b, the matrix ab can be calculated. According to the determinant product rule, the determinant is equal to one, but the trace should also have the correct value t_{ab}. So we also have to demand

$$t_{ab} = \mathrm{Sp}(ab) = AE + BG + CF + DH$$

.

In summary, this means: If, after choosing t_a, t_b, a suitable t_{ab} has been calculated, there are still in $\mathbb{C}^8$ five side conditions to be fulfilled to find suitable a, b. So there should be $8 - 5 = 3$ additional parameters to the free t_a, t_b parameters.

A concrete construction Five nonlinear equations in eight unknowns are not easy to solve. Therefore, we proceed in a different way. First, we define a, b as

$$a = \begin{pmatrix} st_a & Bz_0 \\ C/z_0 & (1-s)t_a \end{pmatrix}, \quad b = \begin{pmatrix} tt_b & F \\ G & (1-t)t_b \end{pmatrix}.$$

Then we proceed as follows:

- The four numbers $s, t, B, F \in \mathbb{C} \setminus \{0\}$ are freely chosen, the value of z_0 is specified later.
- The traces of a or b are then certainly t_a or t_b, and the numbers C and G are determined so that the determinants of a and b are equal to one[17].
- Finally, a value is chosen for z_0, such that the trace of ab is equal to t_{ab}. This results in a quadratic equation for z_0:

$$(BG)z_0^2 + \big(stt_at_b + (1-s)(1-t)t_at_b - t_{ab}z_0 + CF\big) = 0.$$

There is always a solution.

(Seemingly we now have *four* free parameters s, t, B, F, but there should only be three. This is only apparently a contradiction, because by choosing z_0, a three-dimensional surface is defined in the (s, t, B, F)-space.)

You can also do it completely differently by choosing t_a, t_b and determining t_{ab}. This time we start like this:

$$a = \begin{pmatrix} t_a - s & Bz_0 \\ C/z_0 & s \end{pmatrix}, \quad b = \begin{pmatrix} t_b - t & F \\ G & t \end{pmatrix}.$$

---

[17] Explicitly, this means $C = \big(s(1-s)t_a^2 - 1\big)/B$ and $G = \big(t(1-t)t_b^2 - 1\big)/F$.

There s, t, B, F are freely chosen, C, G arise from the condition that the determinants should be one, and z_0 is determined by solving a quadratic equation so that the trace of ab is equal of t_{ab}.

Special cases

Example 1 It starts as usual: You choose any $t_a, t_b \in \mathbb{C}$ and then determine a t_{ab} as the solution of the quadratic equation $x^2 - t_a t_b x + t_a^2 + t_b^2 = 0$. We consider the special case $s = t = 1/2$ in our first example. Then you can find specific values for the matrix entries.

- Define a number z_0 by

$$z_0 = \frac{(t_{ab} - 2)t_b}{t_b t_{ab} - 2t_a + 2t_{ab}}.$$

- Then set

$$a = \begin{pmatrix} \frac{t_a}{2} & \frac{t_a t_{ab} - 2t_b + 4i}{(2t_{ab}+4)z_0} \\ \frac{(t_a t_{ab} - 2t_b - 4i)z_0}{2t_{ab}-4} & \frac{t_a}{2} \end{pmatrix}, \quad b = \begin{pmatrix} \frac{t_b - 2i}{2} & \frac{t_b}{2} \\ \frac{t_b}{2} & \frac{t_b + 2i}{2} \end{pmatrix}.$$

The matrix ab turns out to be remarkably simple:

$$ab = \begin{pmatrix} \frac{t_{ab}}{2} & \frac{t_{ab}-2}{2z_0} \\ \frac{(t_{ab}+2)z_0}{2} & \frac{t_{ab}}{2} \end{pmatrix}.$$

In order to check this, a lot of calculation has to be done. That the determinant of b is equal to one is clear, in the corresponding calculation for a the equation $t_{ab}^2 - t_a t_b t_{ab} + t_a^2 + t_b^2 = 0$ already plays a role, and in the proof of the remaining assertions it is used several times again.

  You can find this "recipe" in the book by Mumford et al. on page 229.

Example 2 In this variant[18] you can also freely choose two complex numbers t_a, t_b: These should later become the traces of a and b. The trace of ab then has to be a solution of the quadratic equation

$$x^2 - t_a t_b x + t_a^2 + t_b^2 = 0$$

because of the Markov identity, if we want to find a, b with the desired properties. Choose such a solution and call it t_{ab}.

---

[18] It is attributed to Jørgensen in the book by Mumford et al.

Suitable matrices a, b are found with the above second general approach by special choice of the free parameters. We set

$$a = \begin{pmatrix} t_a - t_b/t_{ab} & t_a/t_{ab}^2 \\ t_a & t_b/t_{ab} \end{pmatrix}, \quad b = \begin{pmatrix} t_b - t_a/t_{ab} & -t_b/t_{ab}^2 \\ -t_b & t_a/t_{ab} \end{pmatrix}.$$

It is obvious that the traces of a and b have the desired value. That the determinants are equal to one follows from the fact that t_{ab} solves the correct quadratic equation. So, for example, the determinant of a is equal to

$$\left(t_a - \frac{t_b}{t_{ab}}\right)\frac{t_b}{t_{ab}} - \frac{t_a^2}{t_{ab}^2} = \frac{t_b t_a t_{ab} - t_a^2 - t_b^2}{t_{ab}^2},$$

and the numerator is equal to t_{ab}^2 by construction. (For b it can be shown in a similar way)

By repeatedly using the equation $t_{ab}^2 - t_a t_b t_{ab} + t_a^2 + t_b^2 = 0$ it also follows that

$$ab = \begin{pmatrix} t_{ab} & -1/t_{ab} \\ t_{ab} & 0 \end{pmatrix},$$

and therefore ab really has the trace t_{ab}.

The fact that one can now also work with complex entries, unlike in Sect. 6.8, makes the limit sets much more interesting. Here are some examples that were generated with the methods described in the first approach:

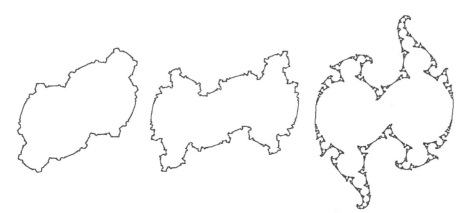

Limits generated with the methods described here

The images of the limit sets generated according to Example 1 show that they are all *point-symmetric*. This is due to the following facts:

Proposition 6.10.2

(i)   Let $\mathcal{G}$ be a discrete group of Möbius transformations with limit set $\mathbb{L}$. If one conjugates $\mathcal{G}$ to $\mathcal{G}_G = \{GMG^{-1} \mid M \in \mathcal{G}\}$, then the following is true: The limit set of $\mathcal{G}_G$ is equal to $G(\mathbb{L})$.

(ii) Let $\mathcal{G}$ be a group generated by transformations a_0, b_0. Further, let G be a transformation such that $Ga_0G^{-1}, Gb_0G^{-1}, Ga_0^{-1}G^{-1}, Gb_0^{-1}G^{-1}$ are in $\mathcal{G}$. Then $G\mathbb{L} = \mathbb{L}$.

(iii) We only assume that Ga_0G^{-1} and Gb_0G^{-1} are in $\mathcal{G}$, and also that $G^{-1} = G$. Even then $G\mathbb{L} = \mathbb{L}$ is true.

Proof (i) follows (for example) from the fact that the limit set is the smallest invariant set. (ii) is true because then $\mathcal{G}$ and $\mathcal{G}_G$ are the same group, and (iii) is obvious.   □

This can be applied here. $\mathcal{G}$ is constructed according to Example 1. $\mathcal{G}$ is also generated by $a_0 := a$ and $b_0 := ab$. The elements on the diagonal of these matrices are identical in each case. If we set $Gz = -z$, then $G^{-1} = G$ and $Ga_0G(z) = a_0^{-1}$ as well as $Gb_0G(z) = b_0^{-1}$ are true. Therefore, $\mathbb{L}$ must be equal to $-\mathbb{L}$.

References for Part II

The following are some selected *references to Part II (Möbius Transformations)*.

Anderson, James W.: Hyperbolic Geometry. Springer, 2005.
This is a good resource for those who want to explore the topic of "Hyperbolic Geometry" in more depth.

Ford, Lester R.: Automorphic Functions. Chelsea Publishers, 1951.
This book provides a good overview of the theory of general Kleinian groups.

Fricke, Robert – Klein, Felix: Vorlesungen über die Theorie der Automorphen Funktionen. Teubner-Verlag, 1897.
A classic text: Here you will find everything that was known about automorphic functions at the end of the 19th century.

Koecher, Max – Krieg, Aloys: Elliptische Funktionen und Modulformen. Springer, 2nd ed., 2007.
This book treats elliptic functions using the theory of discrete groups of Möbius transformations.

Lehner, Josef: Discontinuous Groups and Automorphic Functions. AMS Publications, Mathematical Surveys, 1963.
This book explains why groups of Möbius transformations are important for the theory of complex functions.

D. Mumford, David – Series, Caroline – Wright, David: Indra's Pearls – The Vision of Felix Klein. Cambridge University Press, 2002.
This highly recommended book was the trigger for my intensive study of Kleinian groups.

Needham, Tristan: Visual Complex Analysis. Clarendon Press, 1997.
The theory of complex functions is presented in this book with many illustrations.

Weigert, Elias: Visual Complex Functions. Birkhäuser 2013.
This book also attempts to find new ways to visualize complex functions.

Part III
Penrose Tilings

Penrose Tilings

7

In the present third part of this book we study *Penrose tilings*. So far, groups of movements of the plane have been in the foreground: In the first part they were groups of isometries, in the second part groups of Möbius transformations. In both cases, the concept of the *fundamental domain* played an important role: This was a subset of the plane to which one had to apply all the transformations of the group to tile the plane.

Now the question is what one can say about tilings if certain "basic shapes" $F_1, \ldots, F_k \subset \mathbb{R}^2$ are given. Can one then write the $\mathbb{R}^2$ as $\bigcup_{n \in \mathbb{N}} G_n$, where each G_n results from applying suitable translations, rotations or reflections from one of the $F_1, \ldots, F_k$ and the G_n only intersect at the boundaries, so one speaks of a *tiling*. (A more precise definition is given in the first section of this chapter.) Each of the fundamental domains found in Part 1 gives rise to a tiling with only one basic shape.

A tiling is called *periodic* if there is a non-zero $\mathbf{x}$ such that the decomposition $\mathbb{R}^2 = \bigcup_{n \in \mathbb{N}} G_n$ is invariant under the movement group generated by the translations $T_{\mathbf{x}}$.

For many years it was an *open problem* whether a periodic tiling must exist if there is any tiling at all given $F_1, \ldots, F_k$. The answer is "no". There are many counterexamples, the best known of which date from the seventies of the last century by Roger Penrose. The most important ideas in connection with Penrose tilings will be studied in this chapter.

© The Author(s), under exclusive license to Springer Fachmedien Wiesbaden GmbH, part of Springer Nature 2022
E. Behrends, *Tilings of the Plane*, Mathematics Study Resources 2,
https://doi.org/10.1007/978-3-658-38810-2_7

A Penrose tiling

Penrose tilings are named after the English mathematician and physicist *Roger Penrose* (born 1931). By the way, Penrose met the main character from the first part of this book, Maurits Escher, at the 1954 mathematics congress in Amsterdam. For him Escher's "impossible" figure were particularly interesting, and in collaboration with his father Lionel Penrose he wrote scientific papers about them. (The impossible triangle originally described by Oscar Reutersvard in 1934 is now known as the Penrose triangle.)

His scientific reputation is based primarily on his work on general relativity and cosmology. Penrose tilings were studied in the 1970s, and later a connection was also discovered with (three-dimensional) quasicrystals: these are crystals that only fill space in a non-periodic way.

7.1   Non-periodic Tilings: The Problem

To make the problem described in the introduction more precise, let's imagine a tile layer who has a tile shape or a certain number of different tile shapes available. For example, there could be square tiles, all with the same edge length (one type), or square and triangular tiles, with two triangles making up a square (two types), or …

For each special tile an unlimited number is available, and one may ask whether one can tile the whole plane. However, two special aspects need to be considered. Firstly, unlike with real tiles from the hardware store, it is allowed to turn the tiles over. And secondly, it is required that two tiles meet at an edge that has the same length for both. If this can be achieved, then it is called a *permissible* $(F_1, \ldots, F_k)$ *-tiling*, where the $F_1, \ldots, F_k$ are the tile shapes that can be used.

It is obvious that in the case of "$k = 1$ and $F_1 = $ square" there is a permissible (F_1)-tiling; one just needs to think of an extended chessboard pattern in all directions, where all the fields have the same color. And even with the choice

"$k = 2$ and F_1 = square, F_2 = half square" there are permissible (F_1, F_2)-tilings. Below is an example[1].

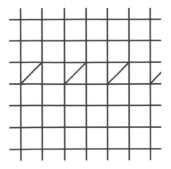

A tiling with two basic shapes

Suppose we have found an admissible $(F_1, \ldots, F_k)$-tiling. Sometimes it is then possible to arrive at the same pattern by shifting it non-trivially. This is the case with our chessboard pattern if one shifts horizontally and vertically by an integer number of square lengths, in the above (F_1, F_2)-tiling shifts by plus or minus 2 square lengths, plus or minus 4 square lengths, ... are admissible in the horizontal direction.

If there is such a shift, the tiling is called *periodic*. It might happen that there are also non-periodic tilings for the tiles $F_1, \ldots, F_k$. This is already the case with our second example. The squares made up of two triangles must only be distributed in the "chessboard" in such a way that the overall pattern cannot be covered by a shift within itself. Something like in the following picture.

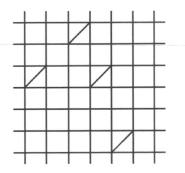

A non-periodic tiling with the same basic shapes

In other words, that means that there can sometimes be both periodic and non-periodic tilings. But it was long open whether this always has to be the case if there are any admissible tilings at all. We formulate the question as

---

[1] One could also lay out the chessboard again, because it is not required that all tile shapes be used.

Classical tiling problem (Hao Wang): Assume there is an admissible $(F_1, \ldots, F_k)$-tiling. Must there then necessarily also exist a periodic tiling?

In 1966, Robert Berger gave an example consisting of 20,526 tiles. A much simpler example was found in 1971 by Raphael Robinson:

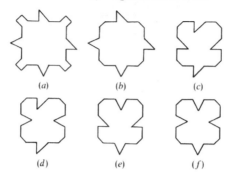

(a)   (b)   (c)

(d)   (e)   (f)

There are only non-periodic tilings for these basic shapes

We will not go into this in more detail in this book. Rather, we will analyze a variant, the *Penrose-triangle tiling*, which was proposed by Roger Penrose in the seventies of the last century. Even there are special tile shapes, but the rules for the creation are modified. And it will turn out:

There are two forms F_1, F_2, so that there are admissible tiling, but all these tiling are aperiodic.

Penrose has dealt several times with the problem of aperiodic tiling. His research was also motivated by physics, because certain crystalline structures (quasicrystals) are aperiodically constructed.

Here we want to deal only with the triangular tiling. (Some references in which one can find the other approaches and more information about aperiodic tiling are summarized in the bibliography.)

Here is the overview of the following sections:

2. The Penrose Tiles
3. If there are Penrose tilings …
4. Index sequences generate tilings
5. Isomorphisms of Penrose tilings
6. Characterization of Penrose tilings
7. Some supplements

It should be emphasized that there are hardly any new results to be found here. The goal is to present the results as completely and as understandable as possible.

7.2   The "Golden" Penrose Triangles

As tiling patterns, only two types of triangles occur, both of which have to do with the *golden ratio*. For the sake of completeness we repeat the definition: One says that two positive numbers a, b are in the ratio of the golden ratio if $a < b$ and

$$\frac{b}{a} = \frac{a+b}{b}$$

hold. It can then be easily proven that $b = \tau a$, where τ is the "golden number" (also "number of the golden ratio")

$$\tau = \frac{1 + \sqrt{5}}{2} = 1.618\ldots$$

τ has many remarkable properties. For example, τ plays an important role in the construction of a pentagon with compasses and ruler, there is an interesting connection to the Fibonacci numbers, etc.

The Penrose triangles of type 1: shapes and colors We start with the following building blocks:

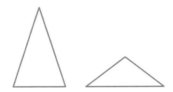

These are two triangles whose sizes are connected with each other as follows:

- The left one is equilateral and has only acute angles. The shortest side has the length a (a number that is arbitrary but fixed), and the other two sides have the length τa. We want to call it $D_g^1(a)$. (The "1" reminds us that we are describing triangles of type 1, and the "g" is the abbreviation of the German word for "large".)
- The right triangle is also equilateral, but has an obtuse angle. The long side has the length $a\tau$, the short sides have the length a. The name: . (Of course, "k" stands for the German word for "small".)

It will be important to be able to distinguish between front and back of these triangles. For this one could attach markings, but we want to make the distinction by colors: $D_g^1(a)$ is blue on one side and dark blue on the other, with $D_k^1(a)$ we use the colors light and dark green. It is agreed that the lighter side is always called the "front" and the darker the "back". So the triangles now look like this from the front (left) and from the back (right):

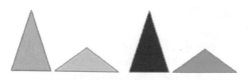

The Penrose triangles of type 1: Laying rules and markings Now the allowed laying rules are to be formulated. We call the sides of $D_g^1(a)$ (seen from the front) in the anti-clockwise A, B and C; where A is the short side.

And the sides of $D_k^1(a)$ are called A', B' and C'; again, the triangle is seen from the front, again the sequence is anti-clockwise, and again we start "at the bottom", this time with the long side.

That looks like this:

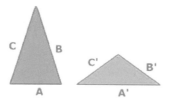

Now let's imagine that for both types of triangles an unlimited number is available, and we want to cover the plane with them without any gaps or overlaps. As before, we demand that two tiles touch at edges of equal length[2].

In addition, there are the **combination rules for triangles of type 1**: You may

- at both types of triangles and at any side continue with a triangle of the same type, if you have previously turned it (i.e. light green to dark green or light blue to dark blue);
- the side A at the side C', if both triangles are seen from the front or both from the back;
- the side B at the side A', if both triangles are seen from the front or both from the back.

And other combination options are *explicitly prohibited*.

---

[2] If we would not prescribe any further conditions, the tiling is not a problem. For example, you could assemble two $D_k^1(a)$-triangles into a parallelogram and then use it to create a tiling quite easily.

Here is an example in which these rules were observed:

In some cases there is only one choice. For example, you can only continue at *B* by turning over. Later we will see that there are more "constraints" built in that you need to consider when you want to tile the whole plane.

However, it is quite laborious if you really want to start tiling and follow all the rules. A help could be *markings* on the sides of the triangle[3]. For example, you could attach small gray semicircles to the sides (on the back, the semicircles should be at the same places) and then require that when attaching, semicircles must always be completed to full circles.

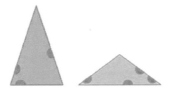

It is not difficult to see that attaching is only allowed when the side markings match.

Penrose solves the problem differently. In this variant, markings in the form of small white and black circle segments are attached at the corners of the triangles. From the front it looks like this (on the back, the same colors are used at corresponding corners):

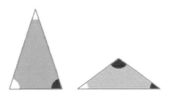

---

[3] The possibility to enforce the conditions as in the case of the above Robinson tiles by suitable serrated triangle sides is ruled out. Because with a serrated edge the combination with a turned tile of the same type will not be possible in all cases that we have in mind,.

The placement rules can be easier to remember if you just follow the *color rule*: You can only combine triangles if the colors at the ends of the common side match. That really covers everything that is stated in the placement rules above. Unfortunately, there is a small *flaw* because *too much is allowed*! The following patterns would also be allowed, but they should not be admissible according to the rules:

To prevent this, the color rule is supplemented by the *parallelogram rule*: Triangles of type $D_g^1(a)$, which can be seen from the front or from the back, must *never* be combined at the white-white edge to form a parallelogram. And the same applies for the black-black edge of $D_k^1(a)$. (Parallelograms that are created by turning in advance are allowed, however.)

We will omit the markings here. For various reasons, it is more reasonable to work directly with the rules described above.

Our *aim* will be to show that one can tile the plane with Penrose triangles in an admissible way and that all these tilings are aperiodic. The proof of these statements is surprisingly difficult. Already for the first part one needs a well thought-out strategy, simply tiling without a strategy is likely to quickly lead to a situation in which it is not possible to continue in an admissible way.

In the *proof* we will need a variant of the tile shapes, the main idea will be to establish relationships between the already defined and the new shapes.

The Penrose triangles of type 2 This is a slightly different situation. There are again two triangles, but they look like this this time:

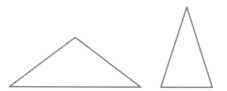

We want to call them $D_g^2(a)$ and $D_k^2(a)$. The first has an obtuse angle, two sides of length τa and one side of length $\tau^2 a$, so it is identical to triangle $D_k^1(\tau a)$. The second is the same as $D_g^1(a)$. There is a front and a back that can be distinguished; here we have colored them light and dark gray (for) or light and dark red (for $D_k^1(a)$). The edges of $D_g^2(a)$ should be labeled as in the following picture with E, F, G, those of $D_k^2(a)$ with E', F', G'. Note: In the picture, the triangles are seen from the front; from the back—then they are dark gray and dark red—the notations F and G as well as the notations F' and G' have to be interchanged.

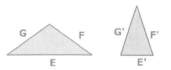

And there are combination rules again, this time the **combination rules for triangles of type 2**. You are allowed to:

- put together triangles only on equal-length sides;
- put together triangles of the same type if one has been turned over over before[4];
- match the side F with the side G' if one of the triangles is seen from the front and the other from the back;
- match the side G with the side F' if both triangles are seen from the front or both from the back.

And more combination possibilities are again *explicitly forbidden*.

7.3   Which Tiling Patterns are Possible?

Now the analysis can begin. For the moment, we forget everything that was said about triangles of type 2 and concentrate on the problem of whether the plane can be tiled with triangles of type 1. The answer is still open, but we want to try to find a solution by "working backwards". To do this, we once assume that we are given a tiling P; whether this could actually happen realistically must remain open for the time being. So we see a lot of triangles of type 1, from the front and from the back. The length of the shorter edge in $D_g^1(a)$, the number a, is actually irrelevant. Nothing will change if we work with $a = 1$, that is, if we see the $D_g^1(1)$ and $D_k^1(1)$ in front of us. As an example, one could think of the pattern at the beginning of the chapter, which we reproduce here once again:

---

[4]So this is exactly as in the case of type 1. Here you are allowed to assemble light gray to dark gray and light red to dark red.

If you look at it long enough, you will notice that not all allowed attachment rules were used. In fact, it would have been admissible for the triangle $D_k^1(1)$ (seen from the front) to be placed with the edge C' on a turned version of $D_k^1(1)$ at B':

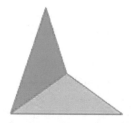

But this does not seem to be the case in the picture. In fact, the following is true:

Lemma 7.3.1

If P is a valid tiling of the plane with triangles of type 1, then no two $D_k^1(1)$ -triangles of different colors (i.e. one seen from the front, one from the back) were ever put together at a C'-edge.

Proof Assume one would find something like this somewhere. The only valid continuation at the G'-edge of the light green triangle would be like this:

But then it can't go any further, because no triangle fits in the gap!   □

Since there were only two possible ways of applying the C'-edge of $D_k^1(1)$, it will always be the other one. In other words: Wherever one sees a $D_k^1(1)$-triangle from the front in the tiling, at the C'-edge will be attached a $D_g^1(1)$-triangle (seen from the front) as its A-edge; and correspondingly, turned $D_k^1(1)$-triangles are always combined at the C'-edge with turned versions of $D_g^1(1)$ at their A-edge.

This has an important consequence. If we imagine such "forced marriages" as new units, then these are $D_g^2(1)$-triangles[5]. Certain $D_g^1(1)$-triangles will remain after this identification. If we consider them as $D_k^1(1)$-triangles, then this means that the plane is now filled seamlessly with type-2-triangles. It is now very important to note that this is also a valid tiling if, when a combined D_g^1-D_k^1-pair is considered, the side "in front" is called the one on which both parts can be seen from the front. And with the new D_k^2-triangles, light blue is again the front side. This can easily be verified, but the result will be proved in more generality in proposition 7.3.3 below. This means:

If there is a valid tiling with triangles of type 1, then there is also one with triangles of type 2.

It is natural to continue this idea. To do this, we look again at the triangles of type 2 and try to find "forced marriages" here as well. There are several of them. For example, it is clear that at the F edge of a $D_g^2(1)$ triangle (seen from the front) only the F edge of a triangle of the same type—but seen from the back—can be attached because all other edges would be too short. But we are interested in a different forced combination. We claim:

Lemma 7.3.2
If P is a valid tiling of the plane with triangles of type 2, then no two $D_k^2(1)$-triangles of different colors (i.e. one seen from the front, one from the back) were ever combined at an F'-edge.

Proof Assume that one could find a constellation like this:

Since there is only one possible combination at the E'-edges, there is only one possible combination to continue:

---

[5] One must must use here the fact that $\tau^2 = \tau + 1$ applies.

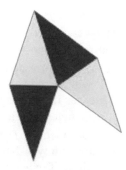

And that is a dead end again, because nothing can be found to continue at the lower gap.   □

This means that one of the two ways in which it could continue with the F' edge is eliminated. So you can be sure that at the F' edge of a $D_k^2(1)$ triangle, a $D_g^2(1)$ triangle with its G edge is always laid, one of the triangles being seen from the front and the other from the back. But such a combination is a $D_g^1(\tau^2)$ triangle, this follows from the equation $\tau^2 + \tau = \tau^3$. In addition, we can notice that $D_g^2(1)$ triangles are also $D_k^1(\tau^2)$ triangles. So if we now consider the pairs of constraints ($D_k^2(1)$ turned over at $D_g^2(1)$, with F' on G) as a new tile, we have again a tiling with Penrose tiles of type 1, the unit of length being increased by the factor τ^2. Also this tiling is—as we will see in a moment—again admissible, if we define "front" and "back" correctly, so that we can repeat the first two steps. In this way we gradually receive coarser and coarser Penrose tilings, which are alternately of type 1 and type 2[6].

In the following picture you can see a section of a tiling and next to it the first coarsening step; in the next step, for example, the two triangles marked by points in the second picture would be combined to form a new large type-1 triangle[7].

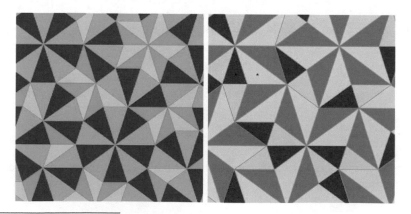

---

[6] The technical term for this process is *deflation*.

[7] Attention: This would then be the view from behind!

In order for the process to never break down, the admissibility of the new tiling must be shown. This is the decisive step in order to be able to completely analyze Penrose tilings.

Proposition 7.3.3

(i) Let Penrose triangles of type 1 be given; we use the above introduced notation. We combine a $D_g^1(a)$-triangle at the A-side with a D_k^1-triangle at the C'-side: This is a $D_g^2(a)$-triangle. It is—per definition—"to be seen from the front" if both sub-triangles are to be seen from the front. Accordingly, we interpret a D_g^1-triangle as a D_k^2-triangle, with the definition of "front" not changing.

Here we see the front sides of the type 2 triangles formed in this way

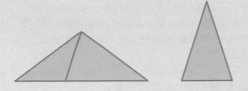

And then the following is true: If one puts these triangles together exactly when the type-1-parts fulfill the type-1-laying rules, then this corresponds to the laying rules of the triangles of type 2 for the newly formed triangles.

(ii) Let Penrose triangles of type 2 be given; we use the same notation as above. We combine a $D_g^2(a)$-triangle seen from the back at the F-side with a $D_k^2(a)$-triangle seen from the front at the G'-side: This is the front side of a $D_g^1(\tau^2 a)$-triangle. Similarly, we think of a D_g^2-triangle as a $D_k^1(\tau^2 a)$-triangle, where the meaning of "front" and "back" does not change. Here we see the front sides of the triangles of type 1 formed in this way.

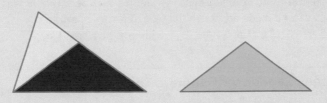

And then the following is true: If you put these triangles together exactly when the type 2-parts satisfy the type 2-laying rules, this corresponds to the laying rules of the triangles of type 1 for the newly formed triangles.

Proof As a reminder, we write the construction rules in a shorter form: "*v*" stands for "different", so one of the triangles should be seen from the front, the other from the back; accordingly, we abbreviate "equal" by "*g* ".

The rules are then:

- Type 1: *A-A-v*, *B-B-v*, *C-C-v*, *A'-A'-v*, *B'-B'-v*, *C'-C'-v*, *A-C'-g*, *B-A'-g*; and everything else is forbidden.
- Type 2: *E-E-v*, *F-F-v*, *G-G-v*, *E'-E'-v*, *F'-F'-v*, *G'-G'-v*, *F-G'-v*, *G-F'-g*; and everything else is forbidden.

It was also proven earlier that when tiling the plane, the actually allowed rules *A'-A'-v* and *F'-F'-v* may not be used if one does not want to end up in a dead end.

(i) In the newly created triangles, *A* and *E'*, *B* and *F'*, *C* and *G'*, *F* and *C*, *G* and *A'* correspond to each other. And *E* has been created by the combination of *B'* and *B*. The "coloring" is as follows: For $D_k^2(a)$, the color of $D_g^1(a)$ is taken over, and $D_g^2(a)$ is exactly seen from the front when both "building blocks" $D_k^1(a)$ and $D_g^1(a)$ are seen from the front.

It is then clear that the construction rules *E-E-v*, *F-F-v*, *G-G-v*, *E'-E'-v*, *F'-F'-v*, and *G'-G'-v* are allowed. From *C-C'-v* it follows that *F-G'-v*, and that *G-F'-g* is admissible, follows from *B-A'-v*.

And everything else is forbidden! *G-G'-g* for example because *A'-A'-g* is taboo. And the inadmissibility *F-G'-g* corresponds to the inadmissibility *C-C'-g*. And so on.

(ii) This proof requires more attention because you have to pay very close attention to the coloring: "front" for the new $D_k^1(\tau^2 a)$ is the same as "front" for the old $D_g^2(a)$, but "front" for the new $D_g^1(\tau^2 a)$ applies when you add a $D_g^2(a)$ seen from the back to a $D_k^2(a)$ seen from the front. Then the desired rules are allowed again. For example *A-C'-g* because of *G-F'-g* and *B-A'-g* because of *E-E-v*.   □

This takes us one decisive step further because with the above considerations we have the opportunity to combine type 1 triangles in a permissible way so that the covered area can be made arbitrarily large. From these "arbitrarily large permissible covered areas" to a tiling of the $\mathbb{R}^2$ it will only be a small step.

7.4   Index Sequences Generate Tilings

It is still not clear whether there are any permissible tilings of the plane. Now we want to read our previous considerations backwards, so to speak, in order to cover the plane by alternately using ever-larger triangles of type 1 and type 2.

The first step: The standard triangles We construct a sequence of pairs of triangles: a pair of type 1, a pair of type 2, a (larger) pair of type 1, a (still larger) pair of type 2, and so on.

In this construction, we remember the "constraints" that we discovered in the previous section: $D_k^1(a)$ is always combined with $D_g^1(a)$ in a very specific way, and the same is true for $D_k^2(a)$, which is always combined with $D_g^2(a)$ in a permissible way.

Now let's get started. We begin with two triangles $D_g^1(1)$ and $D_k^1(1)$ that we see from the front; we want to call them $D(0,0)$ and $D(0,1)$:

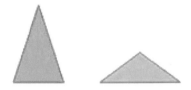

The triangles $D(1,0)$ and $D(1,1)$ then correspond to those that arose in our above analysis after the first "forced marriage". $D(1,1)$ is simply the triangle $D(0,0)$, and $D(1,0)$ is formed by a special combination of the triangles of the previous stage:

And so we continue. $D(2,0)$ is formed by the "forced marriage" of $D(1,0)$ and $D(1,1)$ [8], and $D(2,1)$ is identical with $D(1,0)$. In this way, ever-larger pairs of triangles $D(k,0)$ and $D(k,1)$ are formed, which alternately have type 1 (for even k) and type 2 (for odd k). Here we see more pairs ($k = 2, 3, 4, 7$; the last one is shown reduced):

---

[8] It is important at this point that we do not accidentally interchange front and back: We have to turn $D(1,0)$ over and then apply $D(1,1)$, not the other way around!

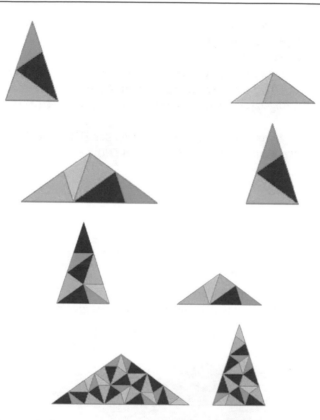

The second step: index sequences By the foregoing construction, it is now possible to create arbitrarily large admissibly covered areas, and one might get the impression that one is well on the way to covering the entire plane. However, there is unfortunately a problem, because when passing from the $D(k,0), D(k,1)$ to the $D(k+1,0), D(k+1,1)$, one not only puts building blocks together, but for odd k, the triangles are also reflected.

But this difficulty is easily remedied. We consider any sequence $(I_k)_{k=0,1,2,...}$ consisting only of zeros and ones (and for which we will soon demand another condition). This could, for example, be the sequence $0, 1, 0, 0, 1, 0, \ldots$.

We mark one of the two standard triangles $D(0,0), D(0,1)$ on both sides with a red point, namely $D(0,0)$ if $I_0 = 0$ and $D(0,1)$ if $I_0 = 1$ Next we consider $D(1,0), D(1,1)$. Actually, there would now be a red point on both triangles, but we only want one. If $D(1,0)$ has the red point, it remains in the case $I_1 = 1$ and the other is deleted, and if $D(1,1)$ has it, the point only survives at $D(1,1)$. Now the procedure should be continued, but there is a peculiarity: If $D(1,1)$ is marked (i.e. in the case $I_1 = 1$), then in the next step only the triangle $D(2,0)$ has a red point. And if then $I_2 = 1$ were admissible, the point would also be deleted and there would nothing be left. This forces an additional condition on the sequence (I_k): A 1 must always be followed by a zero. Then (I_k) is called an *index sequence*. With that we can really mark exactly one triangle in each stage: In the case $I_{k+1} = 0$ the point moves to $D(k + 1, 1)$ and for $I_{k+1} = 0$ to $D(k + 1, 0)$. And the point always sits at the triangle that we had marked at the beginning. Here is an example of markings if we had chosen the index sequence 0, 1, 0, 0:

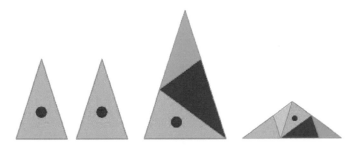

As you can see, the small triangle marked with a growing k must be repeatedly shifted and rotated—and possibly even reflected—in order to cover larger and larger areas of the plane, but in the end we only want to achieve *a tiling* by applying *allowed moves*. This can be easily corrected. We really get *a set of instructions for how to cover arbitrarily large areas:*

- Choose an index sequence $(I_k)_{k=0,...}$.
- Mark $D(0, I_0)$ (at least in your thoughts) with a red dot on both sides.
- Then follow the red dot in the $D(k, 0), D(k, 1)$: The red dot is always at $D(k, I_k)$ in the initially marked triangle.
- Cover larger and larger parts of the plane. It starts with $D(0, I_0)$. And if you have already worked your way up to the k-th step, then you should rotate and fold the marked large triangle $D(k + 1, I_{k+1})$ so that the marked small starting triangle is in the same place and can be seen from the same side as our starting triangle $D(0, I_0)$.
- At the end, the red dot can be removed again. It only were used to give us a helping hand in the construction.

If you really want to do it without the detour via the standard triangles and without red dots, you need a good geometric imagination. The construction steps, where $I_{k+1} = 1$ holds, are trivial, because nothing of importance happens, but in the case of $I_{k+1} = 0$ are for large k in each individual step gigantic triangles to construct, which are composed of many building blocks $D_k^1(1)$ and $D_g^1(1)$.

Here you can see two examples. The upper row shows the beginning of the admissible pattern that belongs to the index sequence $0, 0, 1, 0, \ldots$, and below that, $1, 0, 1, 0, \ldots$ was used.

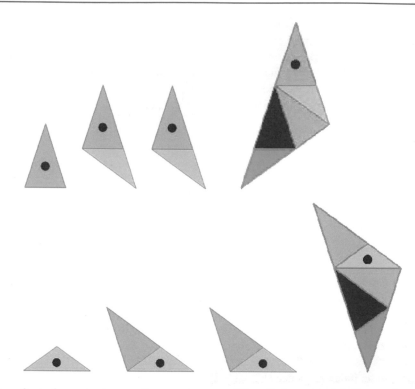

Step 3: Index sequences always generate tilings It is natural to hope that this finally will lead to a tiling of the plane with Penrose triangles of type 1:

Given an index sequence, continue to generate the ever-larger triangles; the union of all these triangles will be called V. If then $V = \mathbb{R}^2$, we have found a tiling.

Unfortunately, there are cases where V is a proper subset of $\mathbb{R}^2$! In the case $(I_k) = (0010001000100010\ldots)$, V is a $36°$ segment, and for the choice $(I_k) = (00000000\ldots)$, V is a half-space. Here we see excerpts, the left pattern continues downward indefinitely, the right one downward to the left:

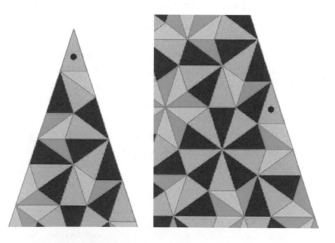

But that's really all that can happen:

> **Proposition 7.4.1**
>
> For V there are only three possibilities:
>
> - $V = \mathbb{R}^2$, or
> - V is a 36°-segment, or
> - V is a half-space.

Proof Suppose V was the whole plane. Then we're done. If that's not the case, there's a point x that's not in V. We start from x in the direction of V until we hit the edge of V^9. $y \in V$ should be the point closest to x, with D we denote a triangle of type 1 that contains y (D is thus a $D_g^1(a)$ or a $D_k^1(a)$).

Now there are several possibilities:

Case 1: y is a corner of D and there's no other triangle of the tiling that contains y. That will then necessarily be a corner with an angle of 36°, because corners with 72° and 108° will reappear during the construction at some point. (For example, the 108° corner in a $D(k, 1)$ is guaranteed to reappear in the next construction step, so at $D_{k+1,0}$.)

We now extend, starting from y, the edges adjacent to y as far as we stay in V. Claim: They can be extended indefinitely. If we were to come to a corner of V again, it could only continue with a kink of 36° due to the previous considerations, and then V would be limited, which is obviously not the case.

In other words: In this case, V is a 36°-segment.

Case 2: At y, the corners of two different small triangles meet. If V had a corner at y, it would be the corner of a $D(k, 0)$ or a $D(k, 1)$, and that would lead using the same argument as before, that only an angle of 36° is possible. But then y could not lie in two triangles, and the only possibility is that y cannot be a corner of V. The edge of V therefore continues in both directions from y.

If it arrives at a corner of V in one direction, we continue the analysis at this corner and conclude that V is a 36°-segment.

Only the case remains that it continues indefinitely in both directions, that is, y lies on a line G and V lies on one side of this line.

If V would not fill this area completely—would therefore V not be a half-space—there would be a point x' there, which does not belong to V. We start our analysis from there again with the choice of a point y' in V as close as possible to x'. The further procedure can then certainly not show that V is a 36°-segment, because such a segment could not contain G as a boundary line. It remains only that y' lies on a boundary line of V. This must necessarily be parallel to G, because otherwise there would be an intersection point and V would have to be a segment.

---

[9] If we want to be a bit more mathematically sound, we need to notice that V as the union of non-overlapping triangles that come in only two versions is necessarily closed. Also, V is convex because in each step we only ever created larger triangles. So there's a point of best approximation to x in V.

But that doesn't work either: V would be the area between two parallel lines in contradiction to the fact that V contains triangles with arbitrarily large diameter. □

Now it is easy to obtain tilings in all cases. If one considers that one can always combine triangles by turning and that—because of the symmetry of the symmetry of the combination conditions—set of tiles admissibly combined also remains admissible when it is turned over, we can conclude our procedure as follows:

• If V is a 36°-segment, reflect it nine times at the apex; this is a tiling with ten 36°-segments.
• If V is a half-space, reflect V at the boundary line to obtain a tiling of the $\mathbb{R}^2$.

We can now be sure that there are admissible Penrose tilings. Our construction procedure even shows more: each index sequence gives rise to such a tiling. The question of whether they are all different will be answered in the next section. But first there is

A supplement: All tilings arise in this way We start again with an admissible tilings P of the plane with type-1 triangles. We fix any triangle D and call it D_0. After coarsening, D lies in a triangle of type 2: That is to say D_1. We continue accordingly: D lies in ever larger triangles of a tessellation, with type 1 and type 2 alternating. The triangle sequence will be called $D_0, D_1, D_2, \ldots$.

Suppose D_0 was a D_g^1 that is seen from the front[10]. We write $\tilde{I}_0 := 0$, and in the case $D_0 = D_k^1$ we would have written $\tilde{I}_0 := 1$. And then we observe in what type of triangle of the first coarsening our D_0 lies. If it was a $D_g^2(1)$, we set $\tilde{I}_1 := 0$, otherwise we define $I_1 := 1$. This continues indefinitely. The (D_k) generate an index sequence, and due to our construction rule, which we have explained in this section, D_k is precisely the triangle that occurs in step k.

Of course, the resulting sequence $(\tilde{I}_k)$ depends on the choice of D_0. For example, consider the 36°-segment shown in proposition 7.4.1. If we choose D_0 as the triangle with the red point, then $(\tilde{I}_k) = (00100010001\ldots)$; this is a special case of the more general statement that the triangle with the point that is supposed to lead to a tiling using (I_k) always leads to $(\tilde{I}_k) = (I_k)$ if we choose it as the starting triangle D_0. But if we decide to choose D_0 as the light blue triangle a little below, then $(\tilde{I}_k) = (0100000001001\ldots)$ results. It is then clear that the index sequences agree from the point on where the triangle with the point and the chosen D_0 are in the same triangle of the growing triangle sequence. We note this important observation as

---

[10] If we have chosen a D_g^1 that is seen from the back, we look at the reflection of P: the tiling is therefore seen from the back. This is "equivalent" to P, more on this in Sect. 5.

Proposition 7.4.2
Let V be the area covered by an index sequence (I_k). Then if we choose any tiling triangle D in V and construct the sequence $(\tilde{I}_k)$ as described above, (I_k) and $(\tilde{I}_k)$ agree from some point on. More precisely: If D is contained in the triangle of the k_0-th construction step, then (I_k) and $(\tilde{I}_k)$ agree at the latest from the point k_0 on.

Back to the tiling P, the triangular sequence (D_k) and the index sequence $(\tilde{I}_k)$, which have resulted from the choice of D_0. We again define V as the union of the D_k. We then already know from proposition 7.4.1 that there are three possibilities for V: an 36°-segment, a half-space or the whole plane. In the latter case, there is nothing further to do, P is then really generated by the index sequence (I_k). But what if V is too small? We know that we can obtain a tiling from V in the first case by nine successive reflections and in the second case by a single reflection, but it is not initially clear that this must coincide with P. That this is indeed the case is the content of

Proposition 7.4.3
We use the above notation and claim:

(i) If V is a 36°-segment, then P is obtained by ninefold reflection at the edges of this segment.

(ii) If V is a half-space, then P is obtained by reflecting V at the boundary line.

Proof We first prove (ii), the boundary line will to be denoted by G. Suppose we could prove the following statement:

Lemma: Let D be a triangle in V that with an edge K touches G. Then P has been extended beyond V in such a way that D is turned over at K[11].

Then it is clear for all edge triangles of V how it continues there with P. But the same result can be applied to larger edge triangles after two coarsenings: They were also turned over, and that already has the consequence for many more initial triangles of V that we find them reflected at G in P. If we repeat this consideration with further coarsenings, we finally exhaust all of V. *Everything* in V is therefore reflected in P. V together with the reflection at G fills the whole plane.

It only remains to prove the *lemma*. The proof is unfortunately a little technical, we have to distinguish six cases. Without loss of generality, let D be seen from the front.

---

[11] In other words, at least at D *must* be turned over to obtain P.

Case 1: D is a D_k^1-triangle that can be seen from the front and touches the C'-edge.

This case is easy to handle because it does not occur: In a tiling of the plane, such a triangle always becomes a D_g^2-triangle in the first coarsening. But that would lead out of V, but should be a subset according to the construction.

Case 2: D is a D_k^1-triangle that can be seen from the front and it touches with *the B'-edge.*

This is also easy because at this edge there is only one allowed attachment rule, namely turning.

Case 3: D is a D_k^1-triangle seen from the front and it touches the A'-edge.

This is a little more difficult. Here on the left in the picture we see D and G (V is thus the upper half-plane), and we want D in P to be turned over. We therefore have to exclude the possibility that the allowed rule B-A'-g was used here, so that it continues on the other side of G as in the right picture:

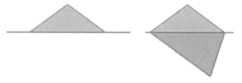

Suppose that this were the case. We claim that the tiling will pass beyond G when continuing to the left of D. We examine P down to the left. A dark blue $D_g^1(1)$ *must* be there. But how should it continue?

Possibility 1: There is a dark green $D_k^1(1)$. This forces—together with the placement constraint at the top left at D—the following configuration, in which the gap at the top left can not be filled in an admissible way.

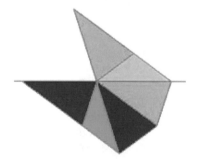

Possibility 2: There is a light blue $D_g^1(1)$. Then you have to put another (then dark blue) $D_g^1(1)$ (next picture to the left), and again we have two choices. You could try a dark green $D_k^1(1)$. But then you have to surpass G, because only at the F' edge can be turned over. Or we can continue with two $D_g^1(1)$ triangles (one light, one dark blue) (right):

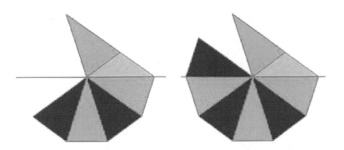

This is also a dead end, because nothing fits in the gap! This proves that the triangle must be mirrored at the A' edge.

Case 4: D is a D_g^1-triangle that can be seen from the front and it touches the C-edge.

This is easy again, because with a C-edge, only mirroring is allowed.

Case 5: D is a D_g^1-triangle that can be seen from the front and touches the A-edge.

Theoretically, there would be two possibilities: mirroring or putting a D_k^1 at the A'-edge. The second is ruled out, because then D would already have landed in a D_g^2-triangle in the first coarsening step, which contains D and therefore cuts into V.

Case 6: D is a D_g^1-triangle that can be seen from the front and touches the B-edge.

Suppose that on the other side there is a D_k^1-triangle $\tilde{D}$ that touches D with the A'-edge. That cannot be the case due to Case 3 if we repeat the same argument from the perspective of $\tilde{D}$.

That completes the proof of (ii).

The statement (i) is now proved quite analogously: The triangles adjacent to the edges of the sectors must be reflected, and by transition to coarser grainings we always obtain more and more necessarily reflected parts of V that successively exhaust V. So on the left and right of the 36°-segment there are reflected versions of V, and if we repeat this a few times, then P is entirely generated by reflections from V[12].   □

This shows that one cannot do it any other way than we did it when generating tiling patterns through index sequences. Or, in other words: All Penrose tilings arise through our process.

It subsequently turns out that there are *three different* classes of these tilings:

- Those for which V is a 36°-segment and the tiling arises by ninefold reflection of V.
- Those for which V is a half-space and only needs to be reflected once.
- Those for which V is already the whole plane.

---

[12] To be precise, one would have to treat the triangle at the tip of the segment separately, but the techniques used so far can be applied also here. In particular, Case 3 of the lemma is important here again. This result guarantees that at the tip of a 36°-segment one has to continue by reflection if the tip is formed by a D_k^1-triangle.

We want to agree for the rest of the work that we call these parquets *ten-fold, two-fold or simple*.

7.5   Isomorphisms of Penrose Tilings

In all areas of mathematics, one must make considerations about an appropriate definition of equality. A geometric construction, for example, is not essentially different if one rotates everything by 30°, and in order to explain what a three-element set is in set theory, one should be able to use $\{1, 2, 3\}$ just as well as $\{10, 20, 30\}$.

Isomorphisms In the context of tilings it is natural to consider situations as "equal", which have arisen from each other by movements of the plane[13]. Two tilings should be called *isomorphic* if they only differ by a movement of the plane.

For example, it is clear that two tilings, which were generated by the same index sequence, but which started at different places on the plane, are isomorphic: By the position of the first triangle and the sequence of indices everything is fixed. The movement, which converts the one start triangle into the other, thus maps the whole first tiling onto the second one.

Symmetries of tilings In this context, the movements of the plane that map a given tiling P onto itself are of interest. One then speaks of *symmetries*. In the constructions of the previous section, we have seen that P may have mirror symmetries (if V is a half-space) or that mirror symmetries and rotational symmetries associated with multiples of 72° are sometimes allowed (if V is a 36°-segment). But we have not yet found that a tiling can be mapped onto itself by displacement. That this really never happens, that therefore *all Penrose tilings are aperiodic*, is part of the following

---

Proposition 7.5.1
Let P be a Penrose tiling and $\phi : \mathbb{R}^2 \to \mathbb{R}^2$ a movement of the plane that transforms P into itself.

(i)   If P is a ten-fold tiling, then ϕ is a rotation by a multiple of 72° around the common apex of the partial segments, possibly followed by a reflection.

(ii) If P is two-fold, then ϕ is the identity or the reflection at the line that bounds the half-space ϕ.

(iii) For simple P, ϕ is necessarily the identity.

---

[13] i.e. by rotations, translations reflections or a combination of these operations.

It follows that the symmetry group of P (which is the collection of all symmetries of P) is always a subgroup of those movements of the plane that transform a regular pentagon into itself. And it also follows that translations can never be symmetries. Or, in other words: All Penrose tilings are aperiodic.

Proof If D is a $D_k^1(1)$-triangle in P, then we are sure that on its C'-edge a $D_g^1(1)$-triangle with its A-edge will be combined and both will be seen from the front or both from the back. At $\phi(D)$, the ϕ-image of the corresponding triangle is thus laid, and that means that ϕ also transforms the first coarsening of P into triangles of type 2 in itself. The same applies to the next coarsening, in which F-G'-ν unions take place and analogously in all further coarsening steps.

We now fix any triangle D in P, call it D_0 and construct the $D_1, D_2, \ldots$ as above in Chap. 4. V should again be the union of the D_k. Denote with $\tilde{D}$ the triangle $\phi(D)$. Analogously, we construct, starting from $\tilde{D}_0 := \tilde{D}$, the ascending triangle sequence $(\tilde{D}_k)$. As a result of our preliminary consideration, we then know that $\tilde{D}_k = \phi(D_k)$ applies.

Suppose $\tilde{D}$ would lie in V. Then $\tilde{D}$ would be contained in k for sufficiently large D_k. On the other hand, since the $\tilde{D}_k$ were constructed solely from the properties of P, $\tilde{D}_k = D_k$ would also be true. And that means that $\phi(D_k) = D_k$ must hold for sufficiently large k. But that is only possible if ϕ is the identity or a reflection at a symmetry axis of D_k. However, since these symmetry axes are different for growing k, the second possibility is ruled out: ϕ is thus the identical mapping. This already proves (iii), because then the condition $\tilde{D} \subset V$ is always fulfilled.

We now prove (i). Since we know the structure of P, we know that we can map $\tilde{D}$ by a suitable movement ψ, which is a rotation by a multiple of $72°$ (possibly followed by a reflection at one of the sector edges), to V. $\psi \circ \phi$ is thus a movement that leaves P invariant, and in addition D is mapped to V. So we know that $\psi \circ \phi$ must be the identity, ψ itself was thus composed of $72°$ rotations and possibly one reflection.

The proof of (ii) is completely analogous, possibly this time you have to mirror everything to fulfill the condition" $\tilde{D} \subset V$ ".   □

How many different Penrose tilings are there? We now know that to each index sequence (I_k) there is a Penrose tiling. There are a lot of them, you can put the "building blocks" "0" and "10" one after the other in any order. This shows that there are uncountably many ways to choose such a sequence. But are the resulting tilings "different"?

In this context, we of course want to interpret "different" as "not isomorphic". The most important result in this context is then the

Proposition 7.5.2
Let P and $\tilde{P}$ be Penrose tilings generated by index sequences (I_k) and $(\tilde{I}_k)$.
Then P and $\tilde{P}$ are isomorphic if and only if the index sequences agree from
some index on.

Proof Let's assume that the index sequences match from the point k_0 on, for
example $I_{k_0} = \tilde{I}_{k_0} = 0$. We recall how we had constructed P: In the k_0-th step,
$D(k_0, 0)$ was rotated and possibly reflected so that the red point came into alignment with the triangle selected at the beginning. The same happened with $\tilde{P}$. And
that's why there is a movement of the plane that brings the k_0-th triangle in the
construction of P into alignment with the one we used in the k_0-th step in the construction of $\tilde{P}$. Since the following indices match, the same movement will also
carry all the following triangles—and thus the union—into itself. Any necessary
steps of rotation and reflection (in the case of tenfold and twofold symmetrical tiling) will then also be carried out at P and $\tilde{P}$ at the same time. This means that both
tilings are isomorphic.

For the second part of the proof, we assume that P and $\tilde{P}$ are isomorphic. The
"starting triangles" shall be denoted by D and $\tilde{D}$, V and $\tilde{V}$ shall be the respective
union of the ascending triangles to be constructed in both situations, and it shall
first be assumed that the isomorphism ϕ maps the triangle D into $\tilde{V}$. If we desig-
nate the partial triangles occurring in the construction with $D_0, D_1, \ldots$ (for P) or
$\tilde{D}_0, \tilde{D}_1, \ldots$ (for $\tilde{P}$), then—since $\tilde{V}$ is the union of $\tilde{D}_k$—the triangle $\hat{D} := \phi(D)$ must
lie in a $\tilde{D}_{k_0}$ if k_0 is only large enough.

We have already noticed in proposition 7.4.2 that this has an important consequence: We can also obtain $\tilde{D}_{k_0}$ by starting with $\phi(D)$. And if we call the resulting
index sequence $(\hat{I}_k)$, then the following holds:

- $\hat{I}_k = I_k$ for all k (cf. the proof of proposition 7.5.1).
- For $k \geq k_0$, $\hat{I}_k = \tilde{I}_k$ (since from k_0 on, the constructed triangles coincide).

If therefore $\phi(D)$ lies in $\tilde{V}$, we are done, because then the I_k and the $\tilde{I}_k$ really coincide from one point on. In the case of ten- or twofold Penrose tilings, we perform
an additional movement if necessary according to ϕ (reflection, rotation by a multiple of 72°). Then P is still mapped to $\tilde{P}$, but we can additionally achieve that D
is mapped to $\tilde{V}$. And since the statement (the index sequences coincide from one
point on) has already been proven for this case, the claim is also shown for such
situations.   □

As an important consequence, we obtain

Proposition 7.5.3
There are uncountably many, non-isomorphic Penrose tilings.

Proof Every index sequence (I_k) generates a tiling $P_{(I_k)}$, and isomorphism occurs precisely when the index sequences agree from one point on. So if we identify two index sequences when the resulting tilings are isomorphic, then the uncountable set *all* of index sequences decomposes into subsets that are each countable: Each subset is formed by the sequences that differ from a fixed sequence at at most finitely many places. But then the number of these subsets must be uncountable, because otherwise the "countable times countable equals countable" theorem would lead to a contradiction. □

7.6 Supplements

Here we collect some results and remarks that are of interest in connection with Penrose tilings.

1. The "golden" triangles used here can all be found in the pentagon. The star formed by the pentagon's diagonals (the "Drudenfuß") was once believed to have magical powers, supposedly able to ward off evil spirits.

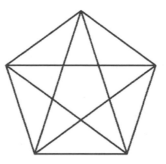

2. Those who would like to experiment with such tilings need "many" $D_g^1(1)$- and $D_k^1(1)$-triangles. What "many" means in each individual case depends on how far in advance one would like to work on the index sequences. For the shopping list, then, the numbers are important. Let's call a_k and b_k the number of large and small triangles, respectively, that one needs to lay the large standard triangle in k-th step. So it is certainly true that $a_0 = 1$ and $b_0 = 0$. The laying instructions provide the recursion equation

$$a_{k+1} = a_k + b_k, \ b_{k+1} = a_k,$$

and it quickly follows that $a_{k+2} = a_{k+1} + a_k$ always holds. The sequence (a_k) therefore begins with

$$1, 1, 2, 3, 5, 8, 13, 21, \ldots$$

This is the sequence of *Fibonacci numbers*, of which one knows that the quotients of two consecutive members converge quite quickly to the golden mean. For us this means: If one would like to buy n D_k^1-triangles, then there should be about $1.62 \cdot n$ D_g^1-triangles, so that it comes out (at least approximately) at the end.

3. Penrose triangles specifically made for this purpose have been used several times at public events since the Year of Mathematics 2008. After an introductory lecture, attendees were able to create some of the ever-growing triangles generated by given index sequences. Here is a picture from "Day of Mathematics 2008" at FU Berlin.

4. We begin with the observation that regardless of the starting triangle and for any choice of the index sequence, in the fourth construction step both sides of both triangles can be seen.

This has an interesting consequence. We imagine any index sequence (I_k) and follow the growing triangles to the k_0-th construction step. Then we see a pattern M. This M will then be a standard triangle $D(k_0, 0)$ (if $I_{k_0} = 0$) or a $D(k_0, 1)$ (if $I_{k_0} = 1$), which may still have to be rotated and / or reflected. This depends on where the red point is.

Now we look at another index sequence $(\tilde{I}_k)$. In the k_0-th step, we also see a rotated or possibly reflected $D(k_0, 0)$ or $D(k_0, 1)$, but after no more than four more steps, the triangle that has arisen contains (possibly up to a rotation) the pattern M due to the preliminary remark. In other words, if one wants to phrase it "philosophically":

Let M be an arbitrarily large, but limited part of a Penrose tiling. Then M is contained in *every* Penrose tiling.

This means that the interesting properties of Penrose tilings only occur when we can control infinitely large areas.

References for Part III

The following are some selected *references to Part III (Penrose tilings).*

Behrends, Ehrhard: Penrose-Parkettierungen mit „goldenen" Dreiecken. Mitt. Math. Ges. Hamburg 31, 2012, 1–28.
The explanations in Chapter 6 are strongly based on this article.

Gardner, Martin: Mathematical Games. Scientific American 1977/1.
This article made Penrose tilings known among mathematicians.

Gardner, Martin: From Penrose Tiles to Trapdoor Ciphers. The Mathematical Association of America, 1997.
Gardner deepens here - without proof - the mathematical background of the Penrose tilings.

Grünbaum, Branko: Tilings and Patterns. W.H. Freeman and Company, 1987.
A classic of tiling theory, which also contains some information on Penrose tilings.

Index

Printed in the United States
by Baker & Taylor Publisher Services